BRIEF ANSWERS
TO THE
BIG QUESTIONS

BRIEF ANSWERS
TO THE
BIG QUESTIONS

STEPHEN
HAWKING

JOHN MURRAY

First published in Great Britain in 2018 by John Murray (Publishers)
An Hachette UK company

This paperback edition published in 2020

20

Copyright © Spacetime Publications Limited 2018
Foreword © Eddie Redmayne 2018
Introduction © Kip S. Thorne 2018
Afterword © Lucy Hawking 2018

A CIP catalogue record for this title is available
from the British Library

Paperback ISBN 978-1-473-69599-3
Ebook ISBN 978-1-473-69600-6

Photograph of the adult Stephen Hawking © Andre Pattenden

Text design by Craig Burgess

Typeset in Sabon MT by Palimpsest Book Production Ltd,
Falkirk, Stirlingshire

Printed and bound in Great Britain by Clays Ltd, Elcograf S.p.A.

John Murray policy is to use papers that are natural, renewable and
recyclable products and made from wood grown in sustainable forests.
The logging and manufacturing processes are expected to conform to
the environmental regulations of the country of origin.

John Murray (Publishers)
Carmelite House
50 Victoria Embankment
London EC4Y 0DZ

www.johnmurraypress.co.uk

Contents

A Note from the Publisher

Stephen Hawking was regularly asked for his thoughts on the 'big questions' of the day by scientists, tech entrepreneurs, senior business figures, political leaders and the general public. Stephen maintained an enormous personal archive of his responses, which took the form of speeches, interviews and essays.

This book draws from this personal archive and was in development at the time of his death. It has been completed in collaboration with his academic colleagues, his family and the Stephen Hawking Estate.

A percentage of the royalties will go to charity.

Foreword

Eddie Redmayne

The first time I met Stephen Hawking, I was struck by his extraordinary power and his vulnerability. The determined look in his eyes coupled with the immobile body was familiar to me from my research – I had recently been engaged to play the role of Stephen in *The Theory of Everything* and had spent several months studying his work and the nature of his disability, attempting to understand how to use my body to express the passage of motor neurone disease over time.

And yet when I finally met Stephen, the icon, this scientist of phenomenal talent, whose main communication was through a computerised voice along with a pair of exceptionally expressive eyebrows, I was floored. I tend to get nervous in silences and talk too much whereas Stephen absolutely understood the power of silence, the power of feeling like you are being scrutinised. Flustered, I chose to talk to him about how our birthdays were only days apart, putting us in the same zodiacal sign. After a few minutes Stephen replied, 'I'm an astronomer. Not an astrologer.' He also insisted that I call him Stephen and stop referring to him as Professor. I had been told . . .

The opportunity to portray Stephen was an extraordinary one. I was drawn to the role because of the duality of Stephen's external triumph in his scientific work and the internal battle against motor neurone disease starting in his early twenties. His was a unique, complex, rich story of human endeavour, family life, huge academic achievement and sheer defiance in the face of all obstacles. While we wanted to portray the inspiration, we also wanted to show the grit and courage involved in Stephen's life, displayed both by him and by those who cared for him.

But it was equally important to portray that side of Stephen which was pure showman. In my trailer I ended up having three images that I referred to. One was Einstein with his tongue out, because there's that similar playful wit with Hawking. Another was the Joker in a pack of cards who's a puppeteer, because I feel Stephen always had people in the palm of his hand. And the third was James Dean. And that was what I gained from seeing him – the glint and the humour.

The greatest pressure in playing a living person is that you will have to account for your performance to the person you have portrayed. In Stephen's case, the accounting was also to his family, who had been so generous to me during my preparation for the film. Before Stephen went into the screening, he said to me, 'I will tell you what I think. Good. Or otherwise.' I replied that if it was 'otherwise' perhaps he could just say 'otherwise' and spare me the excoriating details. Generously, Stephen said he had enjoyed the film. He was moved by it, but famously he also stated that he thought there should have been more physics and fewer feelings. This is impossible to argue with.

Since *The Theory of Everything*, I have stayed in contact with the Hawking family. I was touched to be asked to give a reading at Stephen's funeral. It was an incredibly sad but brilliant day, full of love and joyful memories and reflections on this most courageous of men, who had led the world in his science and in his quest to have disabled people recognised and given proper opportunities to thrive.

We have lost a truly beautiful mind, an astonishing scientist and the funniest man I have ever had the pleasure to meet. But as his family said at the time of Stephen's death, his work and legacy will live on and so it is with sadness but also great pleasure that I introduce you to this collection of Stephen's writings on diverse and fascinating topics. I hope you enjoy his writings and, to quote Barack Obama, I hope Stephen is having fun up there among the stars.

Love
Eddie

An Introduction

Professor Kip S. Thorne

I first met Stephen Hawking in July 1965, in London, England, at a Conference on General Relativity and Gravitation. Stephen was in the midst of his PhD studies at the University of Cambridge; I had just completed mine at Princeton University. Rumours swirled around the conference halls that Stephen had devised a compelling argument that our universe *must* have been born at some finite time in the past. It cannot be infinitely old.

So, along with some 100 people, I squeezed into a room designed for forty, to hear Stephen speak. He walked with a cane and his speech was a bit slurred, but otherwise he showed only modest signs of the motor neurone disease with which he had been diagnosed just two years earlier. His mind was clearly unaffected. His lucid reasoning relied on Einstein's general relativity equations, and on astronomers' observations that our universe is expanding, and on a few simple assumptions that seemed very likely to be true, and it made use of some new mathematical techniques that Roger Penrose had recently devised. Combining all these in ways that were clever, powerful and compelling, Stephen deduced his result: our universe must have begun in some sort of singular state, roughly ten billion years ago. (Over the next decade, Stephen and Roger, combining forces, would go on to prove, ever more convincingly, this singular beginning of time, and also prove ever more convincingly that the core of every black hole is inhabited by a singularity where time ends.)

I emerged from Stephen's 1965 lecture tremendously impressed. Not just by his argument and conclusion, but more importantly by his insightfulness and creativity. So I sought him out and spent an hour talking privately with him. That was the beginning of a lifelong friendship, a friendship based not just on common science interests, but on a remarkable mutual sympathy, an uncanny ability to understand each other as human beings. Soon we were spending more time talking about our lives, our loves, and even death than about science, though our science was still much of the glue that bound us together.

In September 1973 I took Stephen and his wife Jane to Moscow, Russia. Despite the raging Cold War, I had been spending a month or so in Moscow every other year since 1968, collaborating on research with members of a group led by Yakov Borisovich Zel'dovich.

Zel'dovich was a superb astrophysicist, and also a father of the Soviet hydrogen bomb. Because of his nuclear secrets, he was forbidden to travel to Western Europe or America. He craved discussions with Stephen; he could not come to Stephen; so we went to him.

In Moscow, Stephen wowed Zel'dovich and hundreds of other scientists with his insights, and in return Stephen learned a thing or two from Zel'dovich. Most memorable was an afternoon that Stephen and I spent with Zel'dovich and his PhD student Alexei Starobinsky in Stephen's room in the Rossiya Hotel. Zel'dovich explained in intuitive ways a remarkable discovery they had made, and Starobinsky explained it mathematically.

To make a black hole spin requires energy. We already knew that. A black hole, they explained, can use its spin energy to create particles, and the particles will fly away carrying the spin energy with them. This was new and surprising – but not terribly surprising. When an object has energy of motion, nature usually finds a way to extract it. We already knew other ways of extracting a black hole's spin energy; this was just a new, though unexpected way.

Now, the great value of conversations like this is that they can trigger new directions of thought. And so it was with Stephen. He mulled over the Zel'dovich/Starobinsky discovery for several months, looking at it first from one direction and then from another, until one day it triggered a truly radical insight in Stephen's mind: after a black hole stops spinning, the hole can still emit particles. It can radiate – and it radiates as though the black hole was hot, like the Sun, though not very hot, just mildly warm. The heavier the hole, the lower its temperature. A hole that weighs as much as the Sun has a temperature of 0.00000006 Kelvin, 0.06 millionths of a degree above absolute zero. The formula for calculating this temperature is now engraved on Stephen's headstone in Westminster Abbey in London, where his ashes reside between those of Isaac Newton and Charles Darwin.

This 'Hawking temperature' of a black hole and its 'Hawking radiation' (as they came to be called) were truly radical – perhaps the most radical theoretical physics discovery in the second half of the twentieth century. They opened our eyes to profound connections between general relativity (black holes), thermodynamics (the physics of heat) and quantum physics (the creation of particles where before there were none). For example, they led Stephen to prove that a black hole has entropy, which means that somewhere inside or around the black hole there is enormous randomness. He deduced that the amount of entropy (the logarithm of the hole's amount of randomness) is proportional to the hole's surface area. His formula for the entropy is engraved on Stephen's memorial stone at Gonville and Caius College in Cambridge, where he worked.

For the past forty-five years, Stephen and hundreds of other physicists have struggled to understand the precise nature of a black hole's randomness. It is a question that keeps on generating new insights about the marriage of quantum theory with general relativity – that is, about the ill-understood laws of quantum gravity.

In autumn 1974 Stephen brought his PhD students and his family (his wife Jane and their two children Robert and Lucy) to Pasadena, California for a year, so that he and his students could participate in the intellectual life of my university, Caltech, and merge, temporarily, with my own research group. It was a *glorious* year, at the pinnacle of what came to be called 'the golden age of black hole research'.

During that year, Stephen and his students and some of mine struggled to understand black holes more deeply, as did I to some degree. But Stephen's presence, and his leadership in our joint group's black hole research, gave me freedom to pursue a new direction that I had been contemplating for some years: gravitational waves.

There are only two types of waves that can travel across the universe bringing us information about things far away: electromagnetic waves (which include light, X-rays, gamma rays, microwaves, radio waves . . .); and gravitational waves.

Electromagnetic waves consist of oscillating electric and magnetic forces that travel at light speed. When they impinge on charged particles, such as the electrons in a radio or TV antenna, they shake the particles back and forth, depositing in the particles the information the waves carry. That information can then be amplified and fed into a loudspeaker or on to a TV screen for humans to comprehend.

Gravitational waves, according to Einstein, consist of an oscillatory space warp: an oscillating stretch and squeeze of space. In 1972 Rainer (Rai) Weiss at the Massachusetts Institute of Technology had invented a gravitational-wave detector, in which mirrors hanging inside the corner and ends of an L-shaped vacuum pipe are pushed apart along one leg of the L by the stretch of space, and pushed together along the other leg by the squeeze of space. Rai proposed using laser beams to measure the oscillating pattern of this stretch and squeeze. The laser light could extract a gravitational wave's information, and the signal could then be amplified and fed into a computer for human comprehension.

The study of the universe with electromagnetic telescopes (electromagnetic astronomy) was initiated by Galileo, when he built a small optical telescope, pointed it at Jupiter and discovered Jupiter's four largest moons. During the 400 years since then, electromagnetic astronomy has completely revolutionised our understanding of the universe.

In 1972 my students and I began thinking about what

we might learn about the universe using gravitational waves: we began developing a vision for gravitational-wave astronomy. Because gravitational waves are a form of space warp, they are produced most strongly by objects that themselves are made wholly or partially from warped space–time – which means, especially, by black holes. Gravitational waves, we concluded, are the ideal tool for exploring and testing Stephen's insights about black holes.

More generally, it seemed to us, gravitational waves are so radically different from electromagnetic waves that they were almost guaranteed to create their own, new revolution in our understanding of the universe, perhaps comparable to the enormous electromagnetic revolution that followed Galileo – *if* these elusive waves could be detected and monitored. But that was a big *if:* we estimated that the gravitational waves bathing the Earth are so weak that mirrors at the ends of Rai Weiss's L-shaped device would be moved back and forth relative to each other by no more than 1/100th the diameter of a proton (which means 1/10,000,000th of the size of an atom), even if the mirror separation was several kilometres. The challenge of measuring such tiny motions was enormous.

So during that glorious year, with Stephen's and my research groups merged at Caltech, I spent much of my time exploring the prospects for gravitational-wave success. Stephen was helpful in this as, several years earlier, he and his student Gary Gibbons had designed a gravitational-wave detector of their own (which they never built).

Shortly after Stephen's return to Cambridge, my exploration reached fruition with an all-night, intense discussion between Rai Weiss and me in Rai's hotel room in Washington DC. I became convinced that the prospects for success were great enough that I should devote most of my own career, and my future students' research, to helping Rai and other experimenters achieve our gravitational-wave vision. And the rest, as they say, is history.

On 14 September 2015, the LIGO gravitational-wave detectors (built by a 1,000-person project that Rai and I and Ronald Drever co-founded, and Barry Barish organised, assembled and led) registered their first gravitational waves. By comparing the wave patterns with predictions from computer simulations, our team concluded that the waves were produced when two heavy black holes, 1.3 billion light years from Earth, collided. This was the beginning of gravitational-wave astronomy. Our team had achieved, for gravitational waves, what Galileo achieved for electromagnetic waves.

I am confident that, over the coming several decades, the next generation of gravitational-wave astronomers will use these waves not only to test Stephen's laws of black hole physics, but also to detect and monitor gravitational waves from the singular birth of our universe, and thereby test Stephen's and others' ideas about how our universe came to be.

During our glorious year of 1974–5, while I was dithering over gravitational waves, and Stephen was leading our merged group in black hole research, Stephen himself had an insight even more radical than his discovery of Hawking radiation. He gave a compelling, *almost* airtight proof that, when a black hole forms and then subsequently evaporates away completely by emitting radiation, the information that went into the black hole cannot come back out. Information is inevitably lost.

This is radical because the laws of quantum physics insist unequivocally that information can never get totally lost. So, if Stephen was right, black holes violate a most fundamental quantum mechanical law.

How could this be? The black hole's evaporation is governed by the combined laws of quantum mechanics and general relativity – the ill-understood laws of quantum gravity; and so, Stephen reasoned, the fiery marriage of relativity and quantum physics must lead to information destruction.

The great majority of theoretical physicists find this conclusion abhorrent. They are highly sceptical. And so, for forty-four years they have struggled with this so-called information-loss paradox. It is a struggle well worth the effort and anguish that have gone into it, since this paradox is a powerful key for understanding the quantum gravity laws. Stephen himself, in 2003, found a way that information might escape during the hole's evaporation, but that did not quell theorists' struggles. Stephen did not *prove* that the information escapes, so the struggle continues.

In my eulogy for Stephen, at the interment of his ashes at Westminster Abbey, I memorialised that struggle with these words: 'Newton gave us answers. Hawking gave us questions. And Hawking's questions themselves keep on giving, generating breakthroughs decades later. When ultimately we master the quantum gravity laws, and comprehend fully the birth of our universe, it may largely be by standing on the shoulders of Hawking.'

•

Just as our glorious 1974–5 year was only the beginning for my gravitational-wave quest, so it also was just the beginning for Stephen's quest to understand in detail the laws of quantum gravity and what those laws say about the true nature of a black hole's information and randomness, and also about the true nature of our universe's singular birth, and the true nature of the singularities inside black holes – the true nature of the birth and death of time.

These are big questions. Very big.

I have shied away from big questions. I don't have

enough skills, wisdom or self-confidence to tackle them. Stephen, by contrast, was always attracted to big questions, whether they were deeply rooted in his science or not. He *did* have the necessary skills, wisdom and self-confidence.

This book is a compilation of his answers to the big questions, answers on which he was still working at the time of his death.

Stephen's answers to six of the questions are deeply rooted in his science. (Is there a God? How did it all begin? Can we predict the future? What is inside a black hole? Is time travel possible? How do we shape the future?). Here you will find him discussing in depth the issues that I've described briefly in this Introduction, and also much, much more.

His answers to the other four big questions cannot possibly be rooted solidly in his science. (Will we survive on Earth? Is there other intelligent life in the universe? Should we colonise space? Will artificial intelligence outsmart us?) Nevertheless, his answers display deep wisdom and creativity, as we should expect.

I hope you find his answers as stimulating and insightful as do I. Enjoy!

Kip S. Thorne
July 2018

WHY WE MUST ASK THE BIG QUESTIONS

People have always wanted answers to the big questions. Where did we come from? How did the universe begin? What is the meaning and design behind it all? Is there anyone out there? The creation accounts of the past now seem less relevant and credible. They have been replaced by a variety of what can only be called superstitions, ranging from New Age to *Star Trek*. But real science can be far stranger than science fiction, and much more satisfying.

I am a scientist. And a scientist with a deep fascination with physics, cosmology, the universe and the future of humanity. I was brought up by my parents to have an unwavering curiosity and, like my father, to research and try to answer the many questions that science asks us. I have spent my life travelling across the universe, inside my mind. Through theoretical physics,

I have sought to answer some of the great questions. At one point, I thought I would see the end of physics as we know it, but now I think the wonder of discovery will continue long after I am gone. We are close to some of these answers, but we are not there yet.

The problem is, most people believe that real science is too difficult and complicated for them to understand. But I don't think this is the case. To do research on the fundamental laws that govern the universe would require a commitment of time that most people don't have; the world would soon grind to a halt if we all tried to do theoretical physics. But most people can understand and appreciate the basic ideas if they are presented in a clear way without equations, which I believe is possible and which is something I have enjoyed trying to do throughout my life.

It has been a glorious time to be alive and doing research in theoretical physics. Our picture of the universe has changed a great deal in the last fifty years, and I'm happy if I have made a contribution. One of the great revelations of the space age has been the perspective it has given humanity on ourselves. When we see the Earth from space, we see ourselves as a whole. We see the unity, and not the divisions. It is such a simple image with a compelling message; one planet, one human race.

I want to add my voice to those who demand

immediate action on the key challenges for our global community. I hope that going forward, even when I am no longer here, people with power can show creativity, courage and leadership. Let them rise to the challenge of the sustainable development goals, and act, not out of self-interest, but out of common interest. I am very aware of the preciousness of time. Seize the moment. Act now.

•

I have written about my life before but some of my early experiences are worth repeating as I think about my lifelong fascination with the big questions.

I was born exactly 300 years after the death of Galileo, and I would like to think that this coincidence has had a bearing on how my scientific life has turned out. However, I estimate that about 200,000 other babies were also born that day; I don't know whether any of them were later interested in astronomy.

I grew up in a tall, narrow Victorian house in Highgate, London, which my parents had bought very cheaply during the Second World War when everyone thought London was going to be bombed flat. In fact, a V2 rocket landed a few houses away from ours. I was away with my mother and sister at the time, and fortunately my father was not hurt. For years afterwards, there was a large bomb site down the road in which I used to play with my friend Howard. We investigated the results of the explosion with the same curiosity that drove me my whole life.

In 1950, my father's place of work moved to the northern edge of London, to the newly constructed National Institute for Medical Research in Mill Hill, so my family relocated to the cathedral city of St Albans nearby. I was sent to the High School for Girls, which despite its name took boys up to the age of ten. Later I went to St Albans School. I was never more than about halfway up the class – it was a very bright class – but my classmates gave me the nickname Einstein, so presumably they saw signs of something better. When I was twelve, one of my friends bet another friend a bag of sweets that I would never come to anything.

I had six or seven close friends in St Albans, and I remember having long discussions and arguments about everything, from radio-controlled models to religion. One of the big questions we discussed was the origin of the universe, and whether it required a God to create it and set it going. I had heard that light from distant galaxies was shifted towards the red end of the spectrum and this was supposed to indicate that the universe was expanding. But I was sure there must be some other reason for the red shift. Maybe light got tired and more red on its way to us? An essentially unchanging and everlasting universe seemed so much more natural. (It was only years later, after the discovery of the cosmic microwave background about two years into my PhD research, that I realised I had been wrong.)

I was always very interested in how things operated, and I used to take them apart to see how they worked, but I was not so good at putting them back together again. My practical abilities never matched up to my theoretical qualities. My father encouraged my interest in science and was very keen that I should go to Oxford or Cambridge. He himself had gone to University College, Oxford, so he thought I should apply there. At that time, University College had no fellow in mathematics, so I had little option but to try for a scholarship in natural science. I surprised myself by being successful.

The prevailing attitude at Oxford at that time was very anti-work. You were supposed to be brilliant without effort, or to accept your limitations and get a fourth-class degree. I took this as an invitation to do very little. I'm not proud of this, I'm just describing my attitude at the time, shared by most of my fellow students. One result of my illness has been to change all that. When you are faced with the possibility of an early death, it makes you realise that there are lots of things you want to do before your life is over.

Because of my lack of work, I had planned to get

through the final exam by avoiding questions that required any factual knowledge and focus instead on problems in theoretical physics. But I didn't sleep the night before the exam and so I didn't do very well. I was on the borderline between a first- and second-class degree, and I had to be interviewed by the examiners to determine which I should get. In the interview they asked me about my future plans. I replied that I wanted to do research. If they gave me a first, I would go to Cambridge. If I only got a second, I would stay in Oxford. They gave me a first.

In the long vacation following my final exam, the college offered a number of small travel grants. I thought my chances of getting one would be greater the further I proposed to go, so I said I wanted to go to Iran. In the summer of 1962 I set out, taking a train to Istanbul, then on to Erzuerum in eastern Turkey, then to Tabriz, Tehran, Isfahan, Shiraz and Persepolis, the capital of the ancient Persian kings. On my way home, I and my travelling companion, Richard Chiin, were caught in the Bouin-Zahra earthquake, a massive 7.1 Richter quake that killed over 12,000 people. I must have been near the epicentre, but I was unaware of it because I was ill, and in a bus that was bouncing around on the Iranian roads that were then very uneven.

We spent the next several days in Tabriz, while I recovered from severe dysentery and from a broken rib sustained on the bus when I was thrown against the seat in front, still not knowing of the disaster because we didn't speak Farsi. It was not until we reached Istanbul that we learned what had happened. I sent a postcard to my parents, who had been anxiously waiting for ten days, because the last they had heard I was leaving Tehran for the disaster region on the day of the quake. Despite the earthquake, I have many fond memories of my time in Iran. Intense curiosity about the world can put one in harm's way, but for me this was probably the only time in my life that this was true.

I was twenty in October 1962, when I arrived in Cambridge at the department of applied mathematics and theoretical physics. I had applied to work with Fred Hoyle, the most famous British astronomer of the time. I say astronomer, because cosmology then was hardly recognised as a legitimate field. However, Hoyle had enough students already, so to my great disappointment I was assigned to Dennis Sciama, of whom I had not heard. But it was just as well I hadn't been a student of Hoyle, because I would have been drawn into defending his steady-state theory, a task which would have been harder than negotiating Brexit. I began my work by reading old textbooks on general relativity – as ever, drawn to the biggest questions.

As some of you may have seen from the film in which

Eddie Redmayne plays a particularly handsome version of me, in my third year at Oxford I noticed that I seemed to be getting clumsier. I fell over once or twice and couldn't understand why, and I noticed that I could no longer row a sculling boat properly. It became clear something was not quite right, and I was somewhat disgruntled to be told by a doctor at the time to lay off the beer.

The winter after I arrived in Cambridge was very cold. I was home for the Christmas break when my mother persuaded me to go skating on the lake in St Albans, even though I knew I was not up to it. I fell over and had great difficulty getting up again. My mother realised something was wrong and took me to the doctor.

I spent weeks in St Bartholomew's Hospital in London and had many tests. In 1962, the tests were somewhat more primitive than they are now. A muscle sample was taken from my arm, I had electrodes stuck into me and radio-opaque fluid was injected into my spine, which the doctors watched going up and down on X-rays, as the bed was tilted. They never actually told me what was wrong, but I guessed enough to know it was pretty bad, so I didn't want to ask. I had gathered from the doctors' conversations that it, whatever 'it' was, would only get worse, and there was nothing they could do except give me vitamins. In fact, the doctor who performed the tests washed his hands of me and I never saw him again.

At some point I must have learned that the diagnosis was amyotrophic lateral sclerosis (ALS), a type of motor neurone disease, in which the nerve cells of the brain and spinal cord atrophy and then scar or harden. I also learned that people with this disease gradually lose the ability to control their movements, to speak, to eat and eventually to breathe.

My illness seemed to progress rapidly. Understandably, I became depressed and couldn't see the point of continuing to research my PhD, because I didn't know if I would live long enough to finish it. But then the progression slowed down and I had a renewed enthusiasm for my work. After my expectations had been reduced to zero, every new day became a bonus, and I began to appreciate everything I did have. While there's life, there is hope.

And, of course, there was also a young woman called Jane, whom I had met at a party. She was very determined that together we could fight my condition. Her confidence gave me hope. Getting engaged lifted my spirits, and I realised, if we were going to get married, I had to get a job and finish my PhD. And as always, those big questions were driving me. I began to work hard and I enjoyed it.

To support myself during my studies, I applied for

a research fellowship at Gonvillle and Cauis College. To my great surprise, I was elected and have been a fellow of Caius ever since. The fellowship was a turning point in my life. It meant that I could continue my research despite my increasing disability. It also meant that Jane and I could get married, which we did in July 1965. Our first child, Robert, was born after we had been married about two years. Our second child, Lucy, was born about three years later. Our third child, Timothy, would be born in 1979.

As a father, I would try to instill the importance of asking questions, always. My son Tim once told a story in an interview about asking a question which I think at the time he worried was a bit silly. He wanted to know if there were lots of tiny universes dotted around. I told him never to be afraid to come up with an idea or a hypothesis no matter how daft (his words not mine) it might seem.

•

The big question in cosmology in the early 1960s was did the universe have a beginning? Many scientists were instinctively opposed to the idea, because they felt that a point of creation would be a place where science broke down. One would have to appeal to religion and the hand of God to determine how the universe would start off. This was clearly a fundamental question, and it was just what I needed to complete my PhD thesis.

Roger Penrose had shown that once a dying star had contracted to a certain radius, there would inevitably be a singularity, that is a point where space and time came to an end. Surely, I thought, we already knew that nothing could prevent a massive cold star from collapsing under its own gravity until it reached a singularity of infinite density. I realised that similar arguments could be applied to the expansion of the universe. In this case, I could prove there were singularities where space–time had a beginning.

A eureka moment came in 1970, a few days after the birth of my daughter, Lucy. While getting into bed one evening, which my disability made a slow process, I realised that I could apply to black holes the causal structure theory I had developed for singularity theorems. If general relativity is correct and the energy density is positive, the surface area of the event horizon – the boundary of a black hole – has the property that it always increases when additional matter or radiation falls into it. Moreover, if two black holes collide and merge to form a single black hole, the area of the event horizon around the resulting black hole is greater than the sum of the areas of the event horizons around the original black holes.

This was a golden age, in which we solved most of

the major problems in black hole theory even before there was any observational evidence for black holes. In fact, we were so successful with the classical general theory of relativity that I was at a bit of a loose end in 1973 after the publication with George Ellis of our book *The Large Scale Structure of Space–Time*. My work with Penrose had shown that general relativity broke down at singularities, so the obvious next step would be to combine general relativity – the theory of the very large – with quantum theory – the theory of the very small. In particular, I wondered, can one have atoms in which the nucleus is a tiny primordial black hole, formed in the early universe? My investigations revealed a deep and previously unsuspected relationship between gravity and thermodynamics, the science of heat, and resolved a paradox that had been argued over for thirty years without much progress: how could the radiation left over from a shrinking black hole carry all of the information about what made the black hole? I discovered that information is not lost, but it is not returned in a useful way – like burning an encyclopedia but retaining the smoke and ashes.

To answer this, I studied how quantum fields or particles would scatter off a black hole. I was expecting that part of an incident wave would be absorbed, and the remainder scattered. But to my great surprise I found there seemed to be emission from the black hole itself. At first, I thought this must be a mistake in my calculation. But what persuaded me that it was real was that the emission was exactly what was required to identify the area of the horizon with the entropy of a black hole. This entropy, a measure of the disorder of a system, is summed up in this simple formula

$$S = \frac{Akc^3}{4G\hbar}$$

which expresses the entropy in terms of the area of the horizon, and the three fundamental constants of nature, c, the speed of light, G, Newton's constant of gravitation, and $\hbar$, Planck's constant. The emission of this thermal radiation from the black hole is now called Hawking radiation and I'm proud to have discovered it.

In 1974, I was elected a fellow of the Royal Society. This election came as a surprise to members of my department because I was young and only a lowly research assistant. But within three years I had been promoted to professor. My work on black holes had given me hope that we would discover a theory of everything, and that quest for an answer drove me on.

In the same year, my friend Kip Thorne invited me and my young family and a number of others working in general relativity to the California Institute of Technology (Caltech). For the previous four years, I had been using a manual wheelchair as well as a blue electric three-wheeled car, which went at a slow cycling speed, and in which I sometimes illegally carried passengers. When we went to California, we stayed in a Caltech-owned colonial-style house near campus and there I was able to enjoy full-time use of an electric wheelchair for the first time. It gave me a considerable degree of independence, especially as in the United States buildings and sidewalks are much more accessible for the disabled than they are in Britain.

When we returned from Caltech in 1975, I initially felt rather low. Everything seemed so parochial and restricted in Britain compared to the can-do attitude in America. At the time, the landscape was littered with dead trees killed by Dutch elm disease and the country was beset by strikes. However, my mood lifted as I saw success in my work and was elected, in 1979, to the Lucasian Professorship of Mathematics, a post once held by Sir Isaac Newton and Paul Dirac.

During the 1970s, I had been working mainly on black holes, but my interest in cosmology was renewed by the suggestions that the early universe had gone through a period of rapid inflationary expansion in which its size grew at an ever-increasing rate, like the way prices have increased since the UK's Brexit vote. I also spent time working with Jim Hartle, formulating a theory of the universe's birth that we called 'no boundary'.

By the early 1980s, my health continued to worsen,

and I endured prolonged choking fits because my larynx was weakening and was letting food into my lungs as I ate. In 1985, I caught pneumonia on a trip to CERN, the European Organisation for Nuclear Research, in Switzerland. This was a life-altering moment. I was rushed to the Lucerne Cantonal Hospital and put on to a ventilator. The doctors suggested to Jane that things had progressed to the stage where nothing could be done and that they turn off my ventilator to end my life. But Jane refused and had me flown back to Addenbrooke's Hospital in Cambridge by air ambulance.

As you may imagine this was a very difficult time, but thankfully the doctors at Addenbrooke's tried hard to get me back to how I had been before the visit to Switzerland. However, because my larynx was still allowing food and saliva into my lungs, they had to perform a tracheostomy. As most of you will know, a tracheostomy takes away the ability to speak. Your voice is very important. If it is slurred, as mine was, people can think you are mentally deficient and treat you accordingly. Before the tracheostomy my speech was so indistinct that only people who knew me well could understand me. My children were among the few who could do so. For a while after the tracheostomy, the only way I could communicate was to spell out words, letter by letter, by raising my eyebrows when someone pointed to the right letter on a spelling card.

Luckily a computer expert in California named Walt Woltosz heard of my difficulties. He sent me a computer program he had written called Equalizer. This allowed me to select whole words from a series of menus on the computer screen on my wheelchair by pressing a switch in my hand. Over the years since then, the system has developed. Today I use a program called Acat, developed by Intel, which I control by a small sensor in my glasses via my cheek movements. It has a mobile phone, which gives me access to the internet. I can claim to be the most connected person in the world. I have kept the original speech synthesiser I had, however, partly because I haven't heard one with better phrasing, and partly because by now I identify with this voice, despite its American accent.

I first had the idea of writing a popular book about the universe in 1982, around the time of my no-boundary work. I thought I might make a modest amount to help support my children at school and meet the rising costs of my care, but the main reason was that I wanted to explain how far I felt we had come in our understanding of the universe: how we might be near finding a complete theory that would describe the universe and everything in it. Not only is it important to ask questions and find the answers, as a scientist I felt obligated to communicate with the world what we were learning.

Appropriately enough, *A Brief History of Time* was first published on April Fool's Day in 1988. Indeed, the book was originally meant to be called *From the Big Bang to Black Holes: A Short History of Time*. The title was shortened and changed to 'brief', and the rest is history.

I never expected *A Brief History of Time* to do as well as it has. Undoubtedly, the human-interest story of how I have managed to be a theoretical physicist and a bestselling author despite my disabilities has helped. Not everyone may have finished it or understood everything they read, but they at least grappled with one of the big questions of our existence and got the idea that we live in a universe governed by rational laws that, through science, we can discover and understand.

To my colleagues, I'm just another physicist, but to the wider public I became possibly the best-known scientist in the world. This is partly because scientists, apart from Einstein, are not widely known rock stars, and partly because I fit the stereotype of a disabled genius. I can't disguise myself with a wig and dark glasses – the wheelchair gives me away. Being well known and easily recognisable has its pluses and minuses, but the minuses are more than outweighed by the pluses. People seem genuinely pleased to see me. I even had my biggest-ever audience when I opened the Paralympic Games in London in 2012.

•

What was your dream when you were a child, and did it come true?

I wanted to be a great scientist. However, I wasn't a very good student when I was at school, and was rarely more than halfway up my class. My work was untidy, and my handwriting not very good. But I had good friends at school. And we talked about everything and, specifically, the origin of the universe. This is where my dream began, and I am very fortunate that it has come true.

I have led an extraordinary life on this planet, while at the same time travelling across the universe by using my mind and the laws of physics. I have been to the furthest reaches of our galaxy, travelled into a black hole and gone back to the beginning of time. On Earth, I have experienced highs and lows, turbulence and peace, success and suffering. I have been rich and poor, I have been able-bodied and disabled. I have been praised and criticised, but never ignored. I have been enormously privileged, through my work, in being able to contribute to our understanding of the universe. But it would be an empty universe indeed if it were not for the people I love, and who love me. Without them, the wonder of it all would be lost on me.

And at the end of all this, the fact that we humans, who are ourselves mere collections of fundamental particles of nature, have been able to come to an understanding of the laws governing us, and our universe, is a great triumph. I want to share my excitement about these big questions and my enthusiasm about this quest.

One day, I hope we will know the answers to all these questions. But there are other challenges, other big questions on the planet which must be answered, and these will also need a new generation who are interested and engaged, and have an understanding of science. How will we feed an ever-growing population? Provide clean water, generate renewable energy, prevent and cure disease and slow down global climate change? I hope that science and technology will provide the answers to these questions, but it will take people, human beings with knowledge and understanding, to implement these solutions. Let us fight for every woman and every man to have the opportunity to live healthy, secure lives, full of opportunity and love. We are all time travellers, journeying together into the future. But let us work together to make that future a place we want to visit.

Be brave, be curious, be determined, overcome the odds. It can be done.

1

IS THERE A GOD?

Science is increasingly answering questions that used to be the province of religion. Religion was an early attempt to answer the questions we all ask: why are we here, where did we come from? Long ago, the answer was almost always the same: gods made everything. The world was a scary place, so even people as tough as the Vikings believed in supernatural beings to make sense of natural phenomena like lightning, storms or eclipses. Nowadays, science provides better and more consistent answers, but people will always cling to religion, because it gives comfort, and they do not trust or understand science.

A few years ago, *The Times* newspaper ran a headline on the front page which said 'Hawking: God Did Not Create Universe'. The article was illustrated. God was shown in a drawing by Michelangelo, looking thunderous. They printed a photo of me, looking smug. They made it look like a duel between us. But I don't have a grudge against God. I do not want to give the impression that my work is about proving or disproving the existence of God. My work is about finding a rational framework to understand the universe around us.

For centuries, it was believed that disabled people like me were living under a curse that was inflicted by God. Well, I suppose it's possible that I've upset someone up there, but I prefer to think that everything can be explained another way, by the laws of nature. If you believe in science, like I do, you believe that there are certain laws that are always obeyed. If you like, you can say the laws are the work of God, but that is more a definition of God than a proof of his existence. In about 300 BCE, a philosopher called Aristarchus was fascinated by eclipses, especially eclipses of the Moon. He was brave enough to question whether they really were caused by gods. Aristarchus was a true scientific pioneer. He studied the heavens carefully and reached a bold conclusion: he realised the eclipse was really the shadow of the Earth passing over the Moon, and not a divine event. Liberated by this discovery, he was able to work out what was really going on above his head, and draw diagrams that showed the true relationship of the Sun, the Earth and the Moon. From there he reached even more remarkable conclusions. He deduced that the Earth was not the centre of the universe, as everyone had thought, but that it instead orbits the Sun. In fact, understanding this arrangement explains all eclipses. When the Moon casts its shadow on the Earth, that's a solar eclipse. And when the Earth shades the Moon, that's a lunar eclipse. But Aristarchus took it even further. He suggested that stars were not chinks in the floor of heaven, as his contemporaries believed, but that stars were other suns, like ours, only a very long way away. What a stunning realisation it must have been. The universe is a machine governed by principles or laws – laws that can be understood by the human mind.

I believe that the discovery of these laws has been humankind's greatest achievement, for it's these laws of nature – as we now call them – that will tell us whether we need a god to explain the universe at all. The laws of nature are a description of how things actually work in the past, present and future. In tennis, the ball always goes exactly where they say it will. And there are many other laws at work here too. They govern everything that is going on, from how the energy of the shot is produced in the players' muscles to the speed at which the grass grows beneath their feet. But what's really important is that these physical laws, as well as being unchangeable, are universal. They apply not just to the flight of a ball, but to the motion of a planet, and everything else in the universe. Unlike laws made by humans, the laws of nature cannot be broken – that's why they are so powerful and, when seen from a religious standpoint, controversial too.

If you accept, as I do, that the laws of nature are fixed, then it doesn't take long to ask: what role is there for God? This is a big part of the contradiction between science and religion, and although my views have made headlines, it is actually an ancient conflict. One could define God as the embodiment of the laws of nature. However, this is not what most people would think of as God. They mean a human-like being, with whom one can have a personal relationship. When you look at the vast size of the universe, and how insignificant and accidental human life is in it, that seems most implausible.

I use the word 'God' in an impersonal sense, like Einstein did, for the laws of nature, so knowing the mind of God is knowing the laws of nature. My prediction is that we will know the mind of God by the end of this century.

The one remaining area that religion can now lay claim to is the origin of the universe, but even here science is making progress and should soon provide a definitive answer to how the universe began. I published a book that asked if God created the universe, and that caused something of a stir. People got upset that a scientist should have anything to say on the matter of religion. I have no desire to tell anyone what to believe, but for me asking if God exists is a valid question for science. After all, it is hard to think of a more important, or fundamental, mystery than what, or who, created and controls the universe.

I think the universe was spontaneously created out of nothing, according to the laws of science. The basic assumption of science is scientific determinism. The laws of science determine the evolution of the universe, given its state at one time. These laws may, or may not, have been decreed by God, but he cannot intervene to break the laws, or they would not be laws. That leaves God with the freedom to choose the initial state of the universe, but even here it seems there may be laws. So God would have no freedom at all.

Despite the complexity and variety of the universe, it turns out that to make one you need just three ingredients. Let's imagine that we could list them in some kind of cosmic cookbook. So what are the three ingredients we need to cook up a universe? The first is matter – stuff that has mass. Matter is all around us, in the ground beneath our feet and out in space. Dust, rock, ice, liquids. Vast clouds of gas, massive spirals of stars, each containing billions of suns, stretching away for incredible distances.

The second thing you need is energy. Even if you've never thought about it, we all know what energy is. Something we encounter every day. Look up at the Sun and you can feel it on your face: energy produced by a star ninety-three million miles away. Energy permeates the universe, driving the processes that keep it a dynamic, endlessly changing place.

So we have matter and we have energy. The third thing we need to build a universe is space. Lots of space. You can call the universe many things – awesome, beautiful, violent – but one thing you can't call it is cramped. Wherever we look we see space, more space and even more space. Stretching in all directions. It's enough to make your head spin. So where could all this matter, energy and space come from? We had no idea until the twentieth century.

The answer came from the insights of one man, probably the most remarkable scientist who has ever lived. His name was Albert Einstein. Sadly I never got to meet him, since I was only thirteen when he died. Einstein realised something quite extraordinary: that two of the main ingredients needed to make a universe – mass and energy – are basically the same thing, two sides of the same coin if you like. His famous equation $E = mc^2$ simply means that mass can be thought of as a kind of energy, and vice versa. So instead of three ingredients, we can now say that the universe has just two: energy and space. So where did all this energy and space come from? The answer was found after decades of work by scientists: space and energy were spontaneously invented in an event we now call the Big Bang.

At the moment of the Big Bang, an entire universe came into existence, and with it space. It all inflated, just like a balloon being blown up. So where did all this energy and space come from? How does an entire universe full of energy, the awesome vastness of space and everything in it, simply appear out of nothing?

For some, this is where God comes back into the picture. It was God who created the energy and space. The Big Bang was the moment of creation. But science tells a different story. At the risk of getting myself into trouble, I think we can understand much more the natural phenomena that terrified the Vikings. We can even go beyond the beautiful symmetry of energy and matter discovered by Einstein. We can use the laws of nature to address the very origins of the universe, and discover if the existence of God is the only way to explain it.

As I was growing up in England after the Second World War, it was a time of austerity. We were told that you never get something for nothing. But now, after a lifetime of work, I think that actually you can get a whole universe for free.

The great mystery at the heart of the Big Bang is to explain how an entire, fantastically enormous universe of space and energy can materialise out of nothing. The secret lies in one of the strangest facts about our cosmos. The laws of physics demand the existence of something called 'negative energy'.

To help you get your head around this weird but crucial concept, let me draw on a simple analogy. Imagine a man wants to build a hill on a flat piece of land. The hill will represent the universe. To make this hill he digs a hole in the ground and uses that soil to dig his hill. But of course he's not just making a hill – he's also making a hole, in effect a negative version of the hill. The stuff that was in the hole has now become the hill, so it all perfectly balances out. This is the principle behind what happened at the beginning of the universe.

When the Big Bang produced a massive amount of positive energy, it simultaneously produced the same amount of negative energy. In this way, the positive and the negative add up to zero, always. It's another law of nature.

So where is all this negative energy today? It's in the third ingredient in our cosmic cookbook: it's in space. This may sound odd, but according to the laws of nature concerning gravity and motion – laws that are among the oldest in science – space itself is a vast store of negative energy. Enough to ensure that everything adds up to zero.

I'll admit that, unless mathematics is your thing, this is hard to grasp, but it's true. The endless web of billions upon billions of galaxies, each pulling on each other by the force of gravity, acts like a giant storage device. The universe is like an enormous battery storing negative energy. The positive side of things – the mass and energy we see today – is like the hill. The corresponding hole, or negative side of things, is spread throughout space.

So what does this mean in our quest to find out if there is a God? It means that if the universe adds up to nothing, then you don't need a God to create it. The universe is the ultimate free lunch.

Since we know that the positive and the negative add up to zero, all we need to do now is to work out what – or dare I say who – triggered the whole process in the first place. What could cause the spontaneous appearance of a universe? At first, it seems a baffling problem – after all, in our daily lives things don't just materialise out of the blue. You can't just click your fingers and summon up a cup of coffee when you feel like one. You have to make it out of other stuff like coffee beans, water and perhaps some milk and sugar. But travel down into this coffee cup – through the milk particles, down to the atomic level and right down to the sub-atomic level, and you enter a world where conjuring something out of nothing is possible. At least, for a short while. That's because, at this scale, particles such as protons behave according to the laws of nature we call quantum mechanics. And they really can appear at random, stick around for a while and then vanish again, to reappear somewhere else.

Since we know the universe itself was once very small – perhaps smaller than a proton – this means something quite remarkable. It means the universe itself, in all its mind-boggling vastness and complexity, could simply have popped into existence without violating the known laws of nature. From that moment on, vast amounts of energy were released as space itself expanded – a place to store all the negative energy needed to balance the books. But of course the critical question is raised again: did God create the quantum laws that allowed the Big Bang to occur? In a nutshell, do we need a God to set it up so that the Big Bang could bang? I have no desire to offend anyone of faith, but I think science has a more compelling explanation than a divine creator.

Our everyday experience makes us think that everything that happens must be caused by something that occurred earlier in time, so it's natural for us to think that something – maybe God – must have caused the universe to come into existence. But when we're talking about the universe as a whole, that isn't neces-sarily so. Let me explain. Imagine a river, flowing down a mountainside. What caused the river? Well, perhaps the rain that fell earlier in the mountains. But then, what caused the rain? A good answer would be the Sun, that shone down on the ocean and lifted water vapour up into the sky and made clouds. Okay, so what caused the Sun to shine? Well, if we look inside we see the process known as fusion, in which hydrogen atoms join to form helium, releasing vast quantities of energy in the process. So far so good. Where does the hydrogen come from? Answer: the Big Bang. But here's the crucial bit. The laws of nature itself tell us that not only could the universe have popped into existence without any assistance, like a proton, and have required nothing in terms of energy, but also that it is possible that nothing caused the Big Bang. Nothing.

The explanation lies back with the theories of Einstein, and his insights into how space and time in the universe are fundamentally intertwined. Something very wonderful happened to time at the instant of the Big Bang. Time itself began.

To understand this mind-boggling idea, consider a black hole floating in space. A typical black hole is a star so massive that it has collapsed in on itself. It's so massive that not even light can escape its gravity, which is why it's almost perfectly black. It's gravitational pull is so powerful, it warps and distorts not only light but also time. To see how, imagine a clock is being sucked into it. As the clock gets closer and closer to the black

How does God's existence fit into your understanding of the beginning and the end of the universe? And if God was to exist and you had the chance to meet him, what would you ask him?

The question is, 'Is the way the universe began chosen by God for reasons we can't understand, or was it determined by a law of science?' I believe the second. If you like, you can call the laws of science 'God', but it wouldn't be a personal God that you would meet and put questions to. Although, if there were such a God, I would like to ask however did he think of anything as complicated as M-theory in eleven dimensions.

hole, it begins to get slower and slower. Time itself begins to slow down. Now imagine the clock as it enters the black hole – well, assuming of course that it could withstand the extreme gravitational forces – it would actually stop. It stops not because it is broken, but because inside the black hole time itself doesn't exist. And that's exactly what happened at the start of the universe.

In the last hundred years, we have made spectacular advances in our understanding of the universe. We now know the laws that govern what happens in all but the most extreme conditions, like the origin of the universe, or black holes. The role played by time at the beginning of the universe is, I believe, the final key to removing the need for a grand designer and revealing how the universe created itself.

As we travel back in time towards the moment of the Big Bang, the universe gets smaller and smaller and smaller, until it finally comes to a point where the whole universe is a space so small that it is in effect a single infinitesimally small, infinitely dense black hole. And just as with modern-day black holes, floating around in space, the laws of nature dictate something quite extraordinary. They tell us that here too time itself must come to a stop. You can't get to a time before the Big Bang because there was no time before the Big Bang. We have finally found something that doesn't have a cause, because there was no time for a cause to exist in. For me this means that there is no possibility of a creator, because there is no time for a creator to have existed in.

People want answers to the big questions, like why we are here. They don't expect the answers to be easy, so they are prepared to struggle a bit. When people ask me if a God created the universe, I tell them that the question itself makes no sense. Time didn't exist before the Big Bang so there is no time for God to make the universe in. It's like asking for directions to the edge of the Earth – the Earth is a sphere that doesn't have an edge, so looking for it is a futile exercise.

Do I have faith? We are each free to believe what we want, and it's my view that the simplest explanation is that there is no God. No one created the universe and no one directs our fate. This leads me to a profound realisation: there is probably no heaven and afterlife either. I think belief in an afterlife is just wishful thinking. There is no reliable evidence for it, and it flies in the face of everything we know in science. I think that when we die we return to dust. But there's a sense in which we live on, in our influence, and in our genes that we pass on to our children. We have this one life to appreciate the grand design of the universe, and for that I am extremely grateful.

2

HOW DID IT ALL BEGIN?

Hamlet said, 'I could be bounded in a nutshell, and count myself a king of infinite space.' I think what he meant was that although we humans are very limited physically, particularly in my own case, our minds are free to explore the whole universe, and to boldly go where even *Star Trek* fears to tread. Is the universe actually infinite, or just very large? Did it have a beginning? Will it last for ever or just a long time? How can our finite minds comprehend an infinite universe? Isn't it pretentious of us even to make the attempt?

At the risk of incurring the fate of Prometheus, who stole fire from the ancient gods for human use, I believe we can, and should, try to understand the universe. Prometheus' punishment was being chained to a rock for eternity, although happily he was eventually liberated by Hercules. We have already made remarkable progress in understanding the cosmos. We don't yet have a complete picture. I like to think we may not be far off.

According to the Boshongo people of central Africa, in the beginning there was only darkness, water and the great god Bumba. One day Bumba, in pain from stomach ache, vomited up the Sun. The Sun dried up some of the water, leaving land. Still in pain, Bumba vomited up the Moon, the stars and then some animals – the leopard, the crocodile, the turtle and, finally, man.

These creation myths, like many others, try to answer the questions we all ask. Why are we here? Where did we come from? The answer generally given was that humans were of comparatively recent origin because it must have been obvious that the human race was improving its knowledge and technology. So it can't have been around that long or it would have progressed even more. For example, according to Bishop Ussher, the Book of Genesis placed the beginning of time on 22 October 4004 BCE at 6 p.m. On the other hand, the physical surroundings, like mountains and rivers, change very little in a human lifetime. They were therefore thought to be a constant background, and either to have existed for ever as an empty landscape, or to have been created at the same time as the humans.

Not everyone, however, was happy with the idea that the universe had a beginning. For example, Aristotle, the most famous of the Greek philosophers, believed that the universe had existed for ever. Something eternal is more perfect than something created. He suggested the reason we see progress was that floods, or other natural disasters, had repeatedly set civilisation back to the beginning. The motivation for believing in an eternal universe was the desire to avoid invoking divine intervention to create the universe and set it going. Conversely, those who believed that the universe had a beginning used it as an argument for the existence of God as the first cause, or prime mover, of the universe.

If one believed that the universe had a beginning, the obvious questions were, 'What happened before the beginning? What was God doing before he made the world? Was he preparing Hell for people who asked such questions?' The problem of whether or not the universe had a beginning was a great concern to the German philosopher Immanuel Kant. He felt there were logical contradictions, or antinomies, either way. If the universe had a beginning, why did it wait an infinite time before it began? He called that the thesis. On the other hand, if the universe had existed for ever, why did it take an infinite time to reach the present stage? He called that the antithesis. Both the thesis and the antithesis depended on Kant's assumption, along with almost everyone else, that time was absolute. That is to say, it went from the infinite past to the infinite future independently of any universe that might or might not exist.

This is still the picture in the mind of many scientists today. However, in 1915 Einstein introduced his revolutionary general theory of relativity. In this, space and time were no longer absolute, no longer a fixed background to events. Instead, they were dynamical quantities that were shaped by the matter and energy in the universe. They were defined only within the universe, so it made no sense to talk of a time before the universe began. It would be like asking for a point south of the South Pole. It is not defined.

Although Einstein's theory unified time and space, it didn't tell us much about space itself. Something that seems obvious about space is that it goes on and on and on. We don't expect the universe to end in a brick wall, although there's no logical reason why it couldn't. But modern instruments like the Hubble space telescope allow us to probe deep into space. What we see is billions and billions of galaxies, of various shapes and sizes. There are giant elliptical galaxies, and spiral galaxies like our own. Each galaxy contains billions and billions of stars, many of which will have planets round them. Our own galaxy blocks our view in certain directions, but apart from that the galaxies are distributed roughly uniformly throughout space, with some local concentrations and voids. The density of galaxies appears to drop off at very large distances, but that seems to be because they are so far away and faint that we can't make them out. As far as we can tell, the universe goes on in space for ever and is much the same no matter how far it goes on.

Although the universe seems to be much the same at each position in space, it is definitely changing in time. This was not realised until the early years of the last century. Up to then, it was thought the universe was essentially constant in time. It might have existed for an infinite time, but that seemed to lead to absurd conclusions. If stars had been radiating for an infinite time, they would have heated up the universe until it reached their own temperature. Even at night, the whole sky would be as bright as the Sun, because every line of sight would have ended either on a star or on a cloud of dust that had been heated up until it was as hot as the stars. So the observation that we have all made, that the sky at night is dark, is very important. It implies that the universe cannot have existed for ever, in the state we see today. Something must have happened in the past to make the stars turn on a finite time ago. Then the light from very distant stars wouldn't have had time to reach us yet. This would explain why the sky at night isn't glowing in every direction.

If the stars had just been sitting there for ever, why did they suddenly light up a few billion years ago? What was the clock that told them it was time to shine? This puzzled those philosophers, like Immanuel Kant, who believed that the universe had existed for ever. But for most people it was consistent with the idea that the universe had been created, much as it is now, only a few thousand years ago, just as Bishop Ussher had concluded. However, discrepancies in this idea began to appear, with observations by the hundred-inch telescope on Mount Wilson in the 1920s. First of all, Edwin Hubble discovered that many faint patches of light, called nebulae, were in fact other galaxies, vast collections of stars like our Sun, but at a great distance. In order for them to appear so small and faint, the distances had to be so great that light from them would have taken millions or even billions of years to reach us. This indicated that the beginning of the universe couldn't have been just a few thousand years ago.

But the second thing Hubble discovered was even more remarkable. By an analysis of the light from other galaxies, Hubble was able to measure whether they were moving towards us or away. To his great surprise, he found they were nearly all moving away. Moreover, the further they were from us, the faster they were moving away. In other words, the universe is expanding. Galaxies are moving away from each other.

The discovery of the expansion of the universe was one of the great intellectual revolutions of the twentieth century. It came as a total surprise, and it completely changed the discussion of the origin of the universe. If the galaxies are moving apart, they must have been closer together in the past. From the present rate of expansion, we can estimate that they must have been very close together indeed, about 10 to 15 billion years ago. So it looks as though the universe might have started then, with everything being at the same point in space.

But many scientists were unhappy with the universe having a beginning, because it seemed to imply that physics broke down. One would have to invoke an outside agency, which for convenience one can call God, to determine how the universe began. They therefore advanced theories in which the universe was expanding at the present time, but didn't have a beginning. One of these was the steady-state theory, proposed by Hermann Bondi, Thomas Gold and Fred Hoyle in 1948.

In the steady-state theory, as galaxies moved apart, the idea was that new galaxies would form from matter that was supposed to be continually being created throughout space. The universe would have existed for ever, and would have looked the same at all times. This last property had the great virtue of being a definite prediction that could be tested by observation. The Cambridge radio astronomy group, under Martin Ryle, did a survey of weak sources of radio waves in the early

1960s. These were distributed fairly uniformly across the sky, indicating that most of the sources lay outside our galaxy. The weaker sources would be further away, on average.

The steady-state theory predicted a relationship between the number of sources and their strength. But the observations showed more faint sources than predicted, indicating that the density of the sources was higher in the past. This was contrary to the basic assumption of the steady-state theory, that everything was constant in time. For this and other reasons, the steady-state theory was abandoned.

Another attempt to avoid the universe having a beginning was the suggestion that there was a previous contracting phase, but because of rotation and local irregularities the matter would not all fall to the same point. Instead, different parts of the matter would miss each other, and the universe would expand again with the density always remaining finite. Two Russians, Evgeny Lifshitz and Isaak Khalatnikov, actually claimed to have proved that a general contraction without exact symmetry would always lead to a bounce, with the density remaining finite. This result was very convenient for Marxist–Leninist dialectical materialism, because it avoided awkward questions about the creation of the universe. It therefore became an article of faith for Soviet scientists.

I began my research in cosmology just about the time that Lifshitz and Khalatnikov published their conclusion that the universe didn't have a beginning. I realised that this was a very important question, but I wasn't convinced by the arguments that Lifshitz and Khalatnikov had used.

We are used to the idea that events are caused by earlier events, which in turn are caused by still earlier events. There is a chain of causality, stretching back into the past. But suppose this chain has a beginning, suppose there was a first event. What caused it? This was not a question that many scientists wanted to address. They tried to avoid it, either by claiming like the Russians and the steady-state theorists that the universe didn't have a beginning or by maintaining that the origin of the universe did not lie within the realm of science but belonged to metaphysics or religion. In my opinion, this is not a position any true scientist should take. If the laws of science are suspended at the beginning of the universe, might not they also fail at other times? A law is not a law if it only holds sometimes. I believe that we should try to understand the beginning of the universe on the basis of science. It may be a task beyond our powers, but at least we should make the attempt.

Roger Penrose and I managed to prove geometrical theorems to show that the universe must have had a beginning if Einstein's general theory of relativity was correct, and certain reasonable conditions were satisfied. It is difficult to argue with a mathematical theorem, so in the end Lifshitz and Khalatnikov conceded that the universe should have a beginning. Although the idea of a beginning to the universe might not be very welcome to communist ideas, ideology was never allowed to stand in the way of science in physics. Physics was needed for the bomb, and it was important that it worked. However, Soviet ideology did prevent progress in biology by denying the truth of genetics.

Although the theorems Roger Penrose and I proved showed that the universe must have had a beginning, they did not give much information about the nature of that beginning. They indicated that the universe began in a Big Bang, a point where the whole universe and everything in it were scrunched up into a single point of infinite density, a space–time singularity. At this point Einstein's general theory of relativity would have broken down. Thus one cannot use it to predict in what manner the universe began. One is left with the origin of the universe apparently being beyond the scope of science.

Observational evidence to confirm the idea that the universe had a very dense beginning came in October 1965, a few months after my first singularity result, with the discovery of a faint background of microwaves throughout space. These microwaves are the same as those in your microwave oven, but very much less powerful. They would heat your pizza only to minus 270.4 degrees Celsius, not much good for defrosting the pizza, let alone cooking it. You can actually observe these microwaves yourself. Those of you who remember analogue televisions have almost certainly observed these microwaves. If you ever set your television to an empty channel, a few per cent of the snow you saw on the screen was caused by this background of microwaves. The only reasonable interpretation of the background is that it is radiation left over from an early very hot and dense state. As the universe expanded, the radiation would have cooled until it is just the faint remnant we observe today.

That the universe began with a singularity was not an idea that I or a number of other people were happy with. The reason Einstein's general relativity breaks down near the Big Bang is that it is what is called a classical theory. That is, it implicitly assumed what seems obvious from common sense, that each particle had a well-defined position and a well-defined speed. In such a so-called classical theory, if one knows the positions and speeds of all the particles in the universe at one time, one can calculate what they would be at any other time, in the past or future. However, in the early twentieth century scientists discovered that they couldn't calculate exactly what would happen over very short distances. It wasn't just that they needed better theories. There seems to be a certain level of randomness or uncertainty in nature that cannot be removed however good our theories. It can be summed up in the Uncertainty Principle that was proposed in 1927 by the German scientist Werner Heisenberg. One cannot accurately predict both the position and the speed of a particle. The more accurately the position is predicted, the less accurately you will be able to predict the speed, and vice versa.

Einstein objected strongly to the idea that the universe is governed by chance. His feelings were summed up in his dictum 'God does not play dice.' But all the evidence is that God is quite a gambler. The universe is like a giant casino with dice being rolled, or wheels being spun, on every occasion. A casino owner risks losing money each time dice are thrown or the roulette wheel is spun. But over a large number of bets the odds average out, and the casino owner makes sure they average out in his or her favour. That's why casino owners are so rich. The only chance you have of winning against them is to stake all your money on a few rolls of the dice or spins of the wheel.

It is the same with the universe. When the universe is big, there are a very large number of rolls of the dice, and the results average out to something one can predict. But when the universe is very small, near the Big Bang, there are only a small number of rolls of the dice, and the Uncertainty Principle is very important. In order to understand the origin of the universe, one therefore has to incorporate the Uncertainty Principle into Einstein's general theory of relativity. This has been the great challenge in theoretical physics for at least the last thirty years. We haven't solved it yet, but we have made a lot of progress.

Now suppose we try to predict the future. Because we only know some combination of position and speed of a particle, we cannot make precise predictions about the future positions and speeds of particles. We can only assign a probability to particular combinations of positions and speeds. Thus there is a certain probability to a particular future of the universe. But now suppose we try to understand the past in the same way.

Given the nature of the observations we can make now, all we can do is assign a probability to a particular history of the universe. Thus the universe must have many possible histories, each with its own probability. There is a history of the universe in which England win the World Cup again, though maybe the probability is low. This idea that the universe has multiple histories may sound like science fiction, but it is now accepted as science fact. It is due to Richard Feynman, who worked at the eminently respectable California Institute of Technology and played the bongo drums in a strip joint up the road. Feynman's approach to understanding how things works is to assign to each possible history a particular probability, and then use this idea to make predictions. It works spectacularly well to predict the future. So we presume it works to retrodict the past too.

Scientists are now working to combine Einstein's general theory of relativity and Feynman's idea of multiple histories into a complete unified theory that will describe everything that happens in the universe. This unified theory will enable us to calculate how the universe will evolve, if we know its state at one time. But the unified theory will not in itself tell us how the universe began, or what its initial state was. For that, we need something extra. We require what are known as boundary conditions, things that tell us what happens at the frontiers of the universe, the edges of space and time. But if the frontier of the universe was just at a normal point of space and time we could go past it and claim the territory beyond as part of the universe. On the other hand, if the boundary of the universe was at a jagged edge where space or time were scrunched up, and the density was infinite, it would be very difficult to define meaningful boundary conditions. So it is not clear what boundary conditions are needed. It seems there is no logical basis for picking one set of boundary conditions over another.

However, Jim Hartle of the University of California, Santa Barbara, and I realised there was a third possibility. Maybe the universe has no boundary in space and time. At first sight, this seems to be in direct contradiction to the geometrical theorems that I mentioned earlier. These showed that the universe must have had a beginning, a boundary in time. However, in order to make Feynman's techniques mathematically well defined, the mathematicians developed a concept called imaginary time. It isn't anything to do with the real time that we experience. It is a mathematical trick to make the calculations work and it replaces the real time we experience. Our idea was to say that there was no boundary in imaginary time. That did away with trying to invent boundary conditions. We called this the no-boundary proposal.

If the boundary condition of the universe is that it has no boundary in imaginary time, it won't have just a single history. There are many histories in imaginary time and each of them will determine a history in real time. Thus we have a superabundance of histories for the universe. What picks out the particular history, or set of histories that we live in, from the set of all possible histories of the universe?

One point that we can quickly notice is that many

of these possible histories of the universe won't go through the sequence of forming galaxies and stars, something that was essential to our own development. It may be that intelligent beings can evolve without galaxies and stars, but it seems unlikely. Thus the very fact that we exist as beings that can ask the question 'Why is the universe the way it is?' is a restriction on the history we live in. It implies it is one of the minority of histories that have galaxies and stars. This is an example of what is called the Anthropic Principle. The Anthropic Principle says that the universe has to be more or less as we see it, because if it were different there wouldn't be anyone here to observe it.

Many scientists dislike the Anthropic Principle, because it seems little more than hand waving, and not to have much predictive power. But the Anthropic Principle can be given a precise formulation, and it seems to be essential when dealing with the origin of the universe. M-theory, which is our best candidate for a complete unified theory, allows a very large number of possible histories for the universe. Most of these histories are quite unsuitable for the development of intelligent life. Either they are empty, or too short lasting, or too highly curved, or wrong in some other way. Yet, according to Richard Feynman's multiple-histories idea, these uninhabited histories might have quite a high probability.

We really don't care how many histories there may

be that don't contain intelligent beings. We are interested only in the subset of histories in which intelligent life develops. This intelligent life need not be anything like humans. Little green men would do as well. In fact, they might do rather better. The human race does not have a very good record of intelligent behaviour.

As an example of the power of the Anthropic Principle, consider the number of directions in space. It is a matter of common experience that we live in three-dimensional space. That is to say, we can represent the position of a point in space by three numbers. For example, latitude, longitude and height above sea level. But why is space three-dimensional? Why isn't it two, or four, or some other number of dimensions, like in science fiction? In fact, in M-theory space has ten dimensions (as well as the theory having one dimension of time), but it is thought that seven of the ten spatial directions are curled up very small, leaving three directions that are large and nearly flat. It is like a drinking straw. The surface of a straw is two-dimensional. However, one direction is curled up into a small circle, so that from a distance the straw looks like a one-dimensional line.

Why don't we live in a history in which eight of the dimensions are curled up small, leaving only two dimensions that we notice? A two-dimensional animal would have a hard job digesting food. If it had a gut that went right through, like we have, it would divide the animal

What came before the Big Bang?

According to the no-boundary proposal, asking
what came before the Big Bang is meaningless –
like asking what is south of the South Pole
– because there is no notion of time available to
refer to. The concept of time only exists within
our universe.

in two, and the poor creature would fall apart. So two flat directions are not enough for anything as complicated as intelligent life. There is something special about three space dimensions. In three dimensions, planets can have stable orbits around stars. This is a consequence of gravitation obeying the inverse square law, as discovered by Robert Hooke in 1665 and elaborated on by Isaac Newton. Think about the gravitational attraction of two bodies at a particular distance. If that distance is doubled, then the force between them is divided by four. If the distance is tripled then the force is divided by nine, if quadrupled, then the force is divided by sixteen and so on. This leads to stable planetary orbits. Now let's think about four space dimensions. There gravitation would obey an inverse cube law. If the distance between two bodies is doubled, then the gravitational force would be divided by eight, tripled by twenty-seven and if quadrupled, by sixty-four. This change to an inverse cube law prevents planets from having stable orbits around their suns. They would either fall into their sun or escape to the outer darkness and cold. Similarly, the orbits of electrons in atoms would not be stable, so matter as we know it would not exist. Thus although the multiple-histories idea would allow any number of nearly flat directions, only histories with three flat directions will contain intelligent beings. Only in such histories will the question be asked, 'Why does space have three dimensions?'

One remarkable feature of the universe we observe concerns the microwave background discovered by Arno Penzias and Robert Wilson. It is essentially a fossil record of how the universe was when very young. This background is almost the same independently of which direction one looks in. The differences between different directions is about one part in 100,000. These differences are incredibly tiny and need an explanation. The generally accepted explanation for this smoothness is that very early in the history of the universe it underwent a period of very rapid expansion, by a factor of at least a billion billion billion. This process is known as inflation, something that was good for the universe in contrast to inflation of prices that too often plagues us. If that was all there was to it, the microwave radiation would be totally the same in all directions. So where did the small discrepancies come from?

In early 1982, I wrote a paper proposing that these differences arose from the quantum fluctuations during the inflationary period. Quantum fluctuations occur as a consequence of the Uncertainty Principle. Furthermore, these fluctuations were the seeds for structures in our universe: galaxies, stars and us. This idea is basically the same mechanism as so-called Hawking radiation from a black hole horizon, which I had predicted a

decade earlier, except that now it comes from a cosmological horizon, the surface that divided the universe
between the parts that we can see and the parts that we
cannot observe. We held a workshop in Cambridge that
summer, attended by all the major players in the field.
At this meeting, we established most of our present
picture of inflation, including the all-important density
fluctuations, which give rise to galaxy formation and so
to our existence. Several people contributed to the final
answer. This was ten years before fluctuations in the
microwave sky were discovered by the COBE satellite
in 1993, so theory was way ahead of experiment.

Cosmology became a precision science another ten
years later, in 2003, with the first results from the
WMAP satellite. WMAP produced a wonderful map
of the temperature of the cosmic microwave sky, a
snapshot of the universe at just 400,000 years old. The
irregularities you see are predicted by inflation, and
they mean that some regions of the universe had a
slightly higher density than others. The gravitational
attraction of the extra density slows the expansion of
that region, and can eventually cause it to collapse to
form galaxies and stars. So look carefully at the map
of the microwave sky. It is the blueprint for all the
structure in the universe. We are the product of quantum
fluctuations in the very early universe. God really does
play dice.

Superseding WMAP, today there is the Planck satellite, with a much higher-resolution map of the universe. Planck is testing our theories in earnest, and may even detect the imprint of gravitational waves predicted by inflation. This would be quantum gravity written across the sky.

There may be other universes. M-theory predicts that a great many universes were created out of nothing, corresponding to the many different possible histories. Each universe has many possible histories and many possible states as they age to the present and beyond into the future. Most of these states will be quite unlike the universe we observe.

There is still hope that we see the first evidence for M-theory at the LHC particle accelerator, the Large Hadron Collider, at CERN in Geneva. From an M-theory perspective, it only probes low energies, but we might be lucky and see a weaker signal of fundamental theory, such as supersymmetry. I think the discovery of supersymmetric partners for the known particles would revolutionise our understanding of the universe.

In 2012, the discovery of the Higgs particle by the LHC at CERN in Geneva was announced. This was the first discovery of a new elementary particle in the twenty-first century. There is still some hope that the LHC will discover supersymmetry. But even if the LHC does not discover any new elementary particles, supersymmetry might still be found in the next generation of accelerators that are presently being planned.

The beginning of the universe itself in the Hot Big Bang is the ultimate high-energy laboratory for testing M-theory, and our ideas about the building blocks of space–time and matter. Different theories leave behind different fingerprints in the current structure of the universe, so astrophysical data can give us clues about the unification of all the forces of nature. So there may well be other universes, but unfortunately we will never be able to explore them.

We have seen something about the origin of the universe. But that leaves two big questions. Will the universe end? Is the universe unique?

What then will be the future behaviour of the most probable histories of the universe? There seem to be various possibilities, which are compatible with the appearance of intelligent beings. They depend on the amount of matter in the universe. If there is more than a certain critical amount, the gravitational attraction between the galaxies will slow down the expansion.

Eventually they will then start falling towards each other and will all come together in a Big Crunch. That will be the end of the history of the universe, in real time. When I was in the Far East, I was asked not to mention the Big Crunch, because of the effect it might have on the market. But the markets crashed, so maybe the story got out somehow. In Britain, people don't seem too worried about a possible end twenty billion years in the future. You can do quite a lot of eating, drinking and being merry before that.

If the density of the universe is below the critical value, gravity is too weak to stop the galaxies flying apart for ever. All the stars will burn out, and the universe will get emptier and emptier, and colder and colder. So, again, things will come to an end, but in a less dramatic way. Still, we have a few billion years in hand.

In this answer, I have tried to explain something of the origins, future and nature of our universe. The universe in the past was small and dense and so it is quite like the nutshell with which I began. Yet this nut encodes everything that happens in real time. So Hamlet was quite right. We could be bounded in a nutshell and count ourselves kings of infinite space.

3

IS THERE OTHER INTELLIGENT LIFE IN THE UNIVERSE?

I would like to speculate a little on the development of life in the universe, and in particular on the development of intelligent life. I shall take this to include the human race, even though much of its behaviour throughout history has been pretty stupid and not calculated to aid the survival of the species. Two questions I shall discuss are 'What is the probability of life existing elsewhere in the universe?' and 'How may life develop in the future?'

It is a matter of common experience that things get more disordered and chaotic with time. This observation even has its own law, the so-called second law of thermodynamics. This law says that the total amount of disorder, or entropy, in the universe always increases with time. However, the law refers only to the total amount of disorder. The order in one body can increase provided that the amount of *dis*order in its surroundings increases by a greater amount.

This is what happens in a living being. We can define life as an ordered system that can keep itself going against the tendency to disorder and can reproduce itself. That is, it can make similar, but independent, ordered systems. To do these things, the system must convert energy in some ordered form – like food, sunlight or electric power – into disordered energy, in the form of heat. In this way, the system can satisfy the requirement that the total amount of disorder increases while, at the same time, increasing the order in itself and its offspring. This sounds like parents living in a house which gets messier and messier each time they have a new baby.

A living being like you or me usually has two elements: a set of instructions that tell the system how to keep going and how to reproduce itself, and a mechanism to carry out the instructions. In biology, these two parts are called genes and metabolism. But it is worth emphasising that there need be nothing biological about them. For example, a computer virus is a program that will make copies of itself in the memory of a computer, and will transfer itself to other computers. Thus it fits the definition of a living system that I have given. Like a biological virus, it is a rather degenerate form, because it contains only instructions or genes, and doesn't have any metabolism of its own. Instead, it reprograms the metabolism of the host computer, or cell. Some people have questioned whether viruses should count as life, because they are parasites, and cannot exist independently of their hosts. But then most forms of life, ourselves included, are parasites, in that they feed off and depend for their survival on other forms of life. I think computer viruses should count as life. Maybe it says something about human nature that the only form of life we have created so far is purely destructive. Talk about creating life in our own image. I shall return to electronic forms of life later on.

What we normally think of as 'life' is based on chains of carbon atoms, with a few other atoms such as nitrogen or phosphorus. One can speculate that one might have life with some other chemical basis, such as silicon, but carbon seems the most favourable case, because it has the richest chemistry. That carbon atoms should exist at all, with the properties that they have, requires a fine adjustment of physical constants, such as the QCD scale, the electric charge and even the dimension of space–time. If these constants had significantly different values, either the nucleus of the carbon atom would not be stable or the electrons would collapse in on the nucleus. At first sight, it seems remarkable that the universe is so finely tuned. Maybe this is evidence that the universe was specially designed to produce the human race. However, one has to be careful about such arguments, because of the Anthropic Principle, the idea that our theories about the universe must be compatible with our own existence. This is based on the self-evident truth that if the universe had not been suitable for life we wouldn't be asking why it is so finely adjusted. One can apply the Anthropic Principle in either its Strong or Weak versions. For the Strong Anthropic Principle, one supposes that there are many different universes, each with different values of the physical constants. In a small number, the values will allow the existence of objects like carbon atoms, which can act as the building blocks of living systems. Since we must live in one of these universes, we should not be surprised that the physical constants are finely tuned. If they weren't, we wouldn't be here. The Strong form of the Anthropic Principle is thus not very satisfactory, because what operational meaning can one give to the existence of all those other universes? And if they are separate from our own universe, how can what happens in them affect our universe? Instead, I shall adopt what is known as the Weak Anthropic Principle. That is, I shall take the values of the physical constants as given. But I shall see what conclusions can be drawn from the fact that life exists on this planet at this stage in the history of the universe.

There was no carbon when the universe began in the

Big Bang, about 13.8 billion years ago. It was so hot that all the matter would have been in the form of particles called protons and neutrons. There would initially have been equal numbers of protons and neutrons. However, as the universe expanded, it cooled. About a minute after the Big Bang, the temperature would have fallen to about a billion degrees, about a hundred times the temperature in the Sun. At this temperature, neutrons start to decay into more protons.

If this had been all that had happened, all the matter in the universe would have ended up as the simplest element, hydrogen, whose nucleus consists of a single proton. However, some of the neutrons collided with protons and stuck together to form the next simplest element, helium, whose nucleus consists of two protons and two neutrons. But no heavier elements, like carbon or oxygen, would have been formed in the early universe. It is difficult to imagine that one could build a living system out of just hydrogen and helium – and anyway the early universe was still far too hot for atoms to combine into molecules.

The universe continued to expand and cool. But some regions had slightly higher densities than others and the gravitational attraction of the extra matter in those regions slowed down their expansion, and eventually stopped it. Instead, they collapsed to form galaxies and stars, starting from about two billion years after the

Big Bang. Some of the early stars would have been more massive than our Sun; they would have been hotter than the Sun and would have burned the original hydrogen and helium into heavier elements, such as carbon, oxygen and iron. This could have taken only a few hundred million years. After that, some of the stars exploded as supernovae and scattered the heavy elements back into space, to form the raw material for later generations of stars.

Other stars are too far away for us to be able to see directly if they have planets going round them. However, there are two techniques that have enabled us to discover planets around other stars. The first is to look at the star and see if the amount of light coming from it is constant. If a planet moves in front of the star, the light from the star will be slightly obscured. The star will dim a little bit. If this happens regularly, it is because a planet's orbit is taking it in front of the star repeatedly. A second method is to measure the position of the star accurately. If a planet is orbiting the star, it will induce a small wobble in the position of the star. This can be observed and again, if it is a regular wobble, then one deduces that it is due to a planet in orbit around the star. These methods were first applied about twenty years ago and by now a few thousand planets have been discovered orbiting distant stars. It is estimated that one star in five has an Earth-like planet orbiting it at a distance from the star to be compatible with life as we know it. Our own solar system was formed about four and a half billion years ago, or a little more than nine billion years after the Big Bang, from gas contaminated with the remains of earlier stars. The Earth was formed largely out of the heavier elements, including carbon and oxygen. Somehow, some of these atoms came to be arranged in the form of molecules of DNA. This has the famous double-helix form, discovered in the 1950s by Francis Crick and James Watson in a hut on the New Museum site in Cambridge. Linking the two chains in the helix are pairs of nitrogenous bases. There are four types of nitrogenous bases – adenine, cytosine, guanine and thymine. An adenine on one chain is always matched with a thymine on the other chain, and a guanine with a cytosine. Thus the sequence of nitrogenous bases on one chain defines a unique, complementary sequence on the other chain. The two chains can then separate and each acts as a template to build further chains. Thus DNA molecules can reproduce the genetic information coded in their sequences of nitrogenous bases. Sections of the sequence can also be used to make proteins and other chemicals, which can carry out the instructions, coded in the sequence, and assemble the raw material for DNA to reproduce itself.

As I said earlier, we do not know how DNA molecules

first appeared. As the chances against a DNA molecule arising by random fluctuations are very small, some people have suggested that life came to Earth from elsewhere – for instance, brought here on rocks breaking off from Mars while the planets were still unstable – and that there are seeds of life floating round in the galaxy. However, it seems unlikely that DNA could survive for long in the radiation in space.

If the appearance of life on a given planet was very unlikely, one might have expected it to take a long time. More precisely, one might have expected life to appear as late as possible while still allowing time for the subsequent evolution to intelligent beings, like us, before the Sun swells up and engulfs the Earth. The time window in which this could occur is the lifetime of the Sun – about ten billion years. In that time, an intelligent form of life could conceivably master space travel and be able to escape to another star. But if no escape is possible, life on Earth would be doomed.

There is fossil evidence that there was some form of life on Earth about three and a half billion years ago. This may have been only 500 million years after the Earth became stable and cool enough for life to develop. But life could have taken seven billion years to develop in the universe and still have left time to evolve to beings like us, who could ask about the origin of life. If the probability of life developing on a given planet is very small, why did it happen on Earth in about one-fourteenth of the time available?

The early appearance of life on Earth suggests that there is a good chance of the spontaneous generation of life in suitable conditions. Maybe there was some simpler form of organisation which built up DNA. Once DNA appeared, it would have been so successful that it might have completely replaced the earlier forms. We don't know what these earlier forms would have been, but one possibility is RNA.

RNA is like DNA, but rather simpler, and without the double-helix structure. Short lengths of RNA could reproduce themselves like DNA, and might eventually build up to DNA. We cannot make these nucleic acids in the laboratory from non-living material. But given 500 million years, and oceans covering most of the Earth, there might be a reasonable probability of RNA being made by chance.

As DNA reproduced itself, there would have been random errors, many of which would have been harmful and would have died out. Some would have been neutral – they would not have affected the function of the gene. And a few errors would have been favourable to the survival of the species – these would have been chosen by Darwinian natural selection.

The process of biological evolution was very slow at first. It took about two and a half billion years before the earliest cells evolved into multi-cellular organisms. But it took less than another billion years for some of these to evolve into fish, and for some of the fish, in turn, to evolve into mammals. Then evolution seems to have speeded up even more. It took only about a hundred million years to develop from the early mammals to us. The reason is that the early mammals already contained their versions of the essential organs we have. All that was required to evolve from early mammals to humans was a bit of fine-tuning.

But with the human race evolution reached a critical stage, comparable in importance with the development of DNA. This was the development of language, and particularly written language. It meant that information could be passed on from generation to generation, other than genetically through DNA. There has been some detectable change in human DNA, brought about by biological evolution, in the 10,000 years of recorded history, but the amount of knowledge handed on from generation to generation has grown enormously. I have written books to tell you something of what I have learned about the universe in my long career as a scientist, and in doing so I am transferring knowledge from my brain to the page so you can read it.

The DNA in a human egg or sperm contains about three billion base pairs of nitrogenous bases. However, much of the information coded in this sequence seems to be redundant or is inactive. So the total amount of useful information in our genes is probably something like a hundred million bits. One bit of information is the answer to a yes/no question. By contrast, a paperback novel might contain two million bits of information. Therefore, a human's DNA is equivalent to about fifty *Harry Potter* books, and a major national library can contain about five million books – or about ten trillion bits. The amount of information handed down in books or via the internet is 100,000 times as much as there is in DNA.

Even more important is the fact that the information in books can be changed, and updated, much more rapidly. It has taken us several million years to evolve from less advanced, earlier apes. During that time, the useful information in our DNA has probably changed by only a few million bits, so the rate of biological evolution in humans is about a bit a year. By contrast, there are about 50,000 new books published in the English language each year, containing of the order of a hundred billion bits of information. Of course, the great majority of this information is garbage and no use to any form of life. But, even so, the rate at which useful information can be added is millions, if not billions, higher than with DNA.

This means that we have entered a new phase of evolution. At first, evolution proceeded by natural selection – from random mutations. This Darwinian phase lasted about three and a half billion years and produced us, beings who developed language to exchange information. But in the last 10,000 years or so we have been in what might be called an external transmission phase. In this, the *internal* record of information, handed down to succeeding generations in DNA, has changed somewhat. But the *external* record – in books and other long-lasting forms of storage – has grown enormously.

Some people would use the term 'evolution' only for the internally transmitted genetic material and would object to it being applied to information handed down externally. But I think that is too narrow a view. We are more than just our genes. We may be no stronger or inherently more intelligent than our caveman ancestors. But what distinguishes us from them is the knowledge that we have accumulated over the last 10,000 years, and particularly over the last 300. I think it is legitimate to take a broader view and include externally transmitted information, as well as DNA, in the evolution of the human race.

The timescale for evolution in the external transmission period is the timescale for accumulation of information. This used to be hundreds, or even thousands, of years. But now this timescale has shrunk to about fifty years or less. On the other hand, the brains with which we process this information have evolved only on the Darwinian timescale, of hundreds

If intelligent life exists somewhere else than on Earth, would it be similar to the forms we know, or different?

Is there intelligent life on Earth? But seriously, if there is intelligent life elsewhere, it must be a very long way away otherwise it would have visited Earth by now. And I think we would've known if we had been visited; it would be like the film *Independence Day*.

of thousands of years. This is beginning to cause problems. In the eighteenth century, there was said to be a
man who had read every book written. But nowadays,
if you read one book a day, it would take you many
tens of thousands of years to read through the books
in a national library. By which time, many more
books would have been written.

This has meant that no one person can be the master
of more than a small corner of human knowledge.
People have to specialise, in narrower and narrower
fields. This is likely to be a major limitation in the
future. We certainly cannot continue, for long, with
the exponential rate of growth of knowledge that we
have had in the last 300 years. An even greater limitation and danger for future generations is that we still
have the instincts, and in particular the aggressive
impulses, that we had in caveman days. Aggression, in
the form of subjugating or killing other men and taking
their women and food, has had definite survival advantage up to the present time. But now it could destroy
the entire human race and much of the rest of life on
Earth. A nuclear war is still the most immediate danger,
but there are others, such as the release of a genetically
engineered virus. Or the greenhouse effect becoming
unstable.

There is no time to wait for Darwinian evolution
to make us more intelligent and better natured. But we

are now entering a new phase of what might be called self-designed evolution, in which we will be able to change and improve our DNA. We have now mapped DNA, which means we have read 'the book of life', so we can start writing in corrections. At first, these changes will be confined to the repair of genetic defects – like cystic fibrosis and muscular dystrophy, which are controlled by single genes and so are fairly easy to identify and correct. Other qualities, such as intelligence, are probably controlled by a large number of genes, and it will be much more difficult to find them and work out the relations between them. Nevertheless, I am sure that during this century people will discover how to modify both intelligence and instincts like aggression.

Laws will probably be passed against genetic engineering with humans. But some people won't be able to resist the temptation to improve human characteristics, such as size of memory, resistance to disease and length of life. Once such superhumans appear, there are going to be major political problems with the unimproved humans, who won't be able to compete. Presumably, they will die out, or become unimportant. Instead, there will be a race of self-designing beings, who are improving themselves at an ever-increasing rate.

If the human race manages to redesign itself, to reduce or eliminate the risk of self-destruction, it will probably spread out and colonise other planets and stars. However, long-distance space travel will be difficult for chemically based life forms – like us – based on DNA. The natural lifetime for such beings is short compared with the travel time. According to the theory of relativity, nothing can travel faster than light, so a round trip from us to the nearest star would take at least eight years, and to the centre of the galaxy about 50,000 years. In science fiction, they overcome this difficulty by space warps, or travel through extra dimensions. But I don't think these will ever be possible, no matter how intelligent life becomes. In the theory of relativity, if one can travel faster than light, one can also travel back in time, and this would lead to problems with people going back and changing the past. One would also expect to have already seen large numbers of tourists from the future, curious to look at our quaint, old-fashioned ways.

It might be possible to use genetic engineering to make DNA-based life survive indefinitely, or at least for 100,000 years. But an easier way, which is almost within our capabilities already, would be to send machines. These could be designed to last long enough for interstellar travel. When they arrived at a new star, they could land on a suitable planet and mine material to produce more machines, which could be sent on to yet more stars. These machines would be a new form of life, based on mechanical and electronic components rather than macromolecules. They could eventually replace DNA-based life, just as DNA may have replaced an earlier form of life.

•

What are the chances that we will encounter some alien form of life as we explore the galaxy? If the argument about the timescale for the appearance of life on Earth is correct, there ought to be many other stars whose planets have life on them. Some of these stellar systems could have formed five billion years before the Earth – so why is the galaxy not crawling with self-designing mechanical or biological life forms? Why hasn't the Earth been visited and even colonised? By the way, I discount suggestions that UFOs contain beings from outer space, as I think that any visits by aliens would be much more obvious – and probably also much more unpleasant.

So why haven't we been visited? Maybe the probability of life spontaneously appearing is so low that Earth is the only planet in the galaxy – or in the observable universe – on which it happened. Another possibility is that there was a reasonable probability of forming self-reproducing systems, like cells, but that most of these forms of life did not evolve intelligence. We are used to thinking of intelligent life as an inevitable consequence of evolution, but what if it isn't? The Anthropic Principle should warn us to be wary of such arguments. It is more likely that evolution is a random process, with intelligence as only one of a large number of possible outcomes.

It is not even clear that intelligence has any long-term survival value. Bacteria, and other single-cell organisms, may live on if all other life on Earth is wiped out by our actions. Perhaps intelligence was an unlikely development for life on Earth, from the chronology of evolution, as it took a very long time – two and a half billion years – to go from single cells to multi-cellular beings, which are a necessary precursor to intelligence. This is a good fraction of the total time available before the Sun blows up, so it would be consistent with the hypothesis that the probability for life to develop intelligence is low. In this case, we might expect to find many other life forms in the galaxy, but we are unlikely to find intelligent life.

Another way in which life could fail to develop to an intelligent stage would be if an asteroid or comet were to collide with the planet. In 1994, we observed the collision of a comet, Shoemaker–Levy, with Jupiter. It produced a series of enormous fireballs. It is thought the collision of a rather smaller body with the Earth, about sixty-six million years ago, was responsible for the extinction of the dinosaurs. A few small early mammals survived, but anything as large as a human would have almost certainly been wiped out. It is difficult to say how often such collisions occur, but a reasonable guess might be every twenty million years, on average. If this figure is correct, it would mean that intelligent life on Earth has developed only because of the lucky chance that there have been no major collisions in the last sixty-six million years. Other planets in the galaxy, on which life has developed, may not have had a long enough collision-free period to evolve intelligent beings.

A third possibility is that there is a reasonable probability for life to form and to evolve to intelligent beings, but the system becomes unstable and the intelligent life destroys itself. This would be a very pessimistic conclusion and I very much hope it isn't true.

I prefer a fourth possibility: that there are other forms of intelligent life out there, but that we have been overlooked. In 2015 I was involved in the launch of the Breakthrough Listen Initiatives. Breakthrough Listen uses radio observations to search for intelligent extraterrestrial life, and has state-of-the-art facilities, generous funding and thousands of hours of dedicated radio telescope time. It is the largest ever scientific research programme aimed at finding evidence of civilisations beyond Earth. Breakthrough Message is an international competition to create messages that could be read by an advanced civilisation. But we need to be wary of answering back until we have developed a bit further. Meeting a more advanced civilisation, at our present stage, might be a bit like the original inhabitants of America meeting Columbus – and I don't think they thought they were better off for it.

4

CAN WE PREDICT
THE FUTURE?

In ancient times, the world must have seemed pretty arbitrary. Disasters such as floods, plagues, earthquakes or volcanoes must have seemed to happen without warning or apparent reason. Primitive people attributed such natural phenomena to a pantheon of gods and goddesses, who behaved in a capricious and whimsical way. There was no way to predict what they would do, and the only hope was to win favour by gifts or actions. Many people still partially subscribe to this belief and try to make a pact with fortune. They offer to behave better or be kinder if only they can get an A-grade for a course or pass their driving test.

Gradually however, people must have noticed certain regularities in the behaviour of nature. These regularities were most obvious in the motion of the heavenly bodies across the sky. So astronomy was the first science to be developed. It was put on a firm mathematical basis by Newton more than 300 years ago, and we still use his theory of gravity to predict the motion of almost all celestial bodies. Following the example of astronomy, it was found that other natural phenomena also obeyed definite scientific laws. This led to the idea of scientific determinism, which seems first to have been publicly expressed by the French scientist Pierre-Simon Laplace. I would like to quote to you Laplace's actual words, but Laplace was rather like Proust in that he wrote sentences of inordinate length and complexity. So I have decided to paraphrase the quotation. In effect what he said was that if at one time we knew the positions and speeds of all the particles in the universe, then we would be able to calculate their behaviour at any other time in the past or future. There is a probably apocryphal story that when Laplace was asked by Napoleon how God fitted into this system, he replied, 'Sire, I have not needed that hypothesis.' I don't think that Laplace was claiming that God didn't exist. It is just that God doesn't intervene to break the laws of science. That must be the position of every scientist. A scientific law is not a scientific law if it only holds when some supernatural being decides to let things run and not intervene.

The idea that the state of the universe at one time determines the state at all other times has been a central tenet of science ever since Laplace's time. It implies that we can predict the future, in principle at least. In practice, however, our ability to predict the future is severely limited by the complexity of the equations, and the fact that they often have a property called chaos. As those who have seen *Jurassic Park* will know, this means a tiny disturbance in one place can cause a major change in another. A butterfly flapping its wings in Australia can cause rain in Central Park, New York. The trouble is, it is not repeatable. The next time the butterfly flaps its wings a host of other things will be different, which will also influence the weather. This chaos factor is why weather forecasts can be so unreliable.

Despite these practical difficulties, scientific determinism remained the official dogma throughout the nineteenth century. However, in the twentieth century there were two developments that show that Laplace's vision, of a complete prediction of the future, cannot be realised. The first of these developments was what is called quantum mechanics. This was put forward in 1900 by the German physicist Max Planck as an ad hoc hypothesis, to solve an outstanding paradox. According to the classical nineteenth-century ideas dating back to Laplace, a hot body, like a piece of red-hot metal, should give off radiation. It would lose energy in radio waves, the infra-red, visible light, ultra-violet, X-rays and gamma rays, all at the same rate.

This would mean not only that we would all die of skin cancer, but also that everything in the universe would be at the same temperature, which clearly it isn't.

However, Planck showed one could avoid this disaster if one gave up the idea that the amount of radiation could have just any value, and said instead that radiation came only in packets or quanta of a certain size. It is a bit like saying that you can't buy sugar loose in the supermarket, it has to be in kilogram bags. The energy in the packets or quanta is higher for ultra-violet and X-rays than for infra-red or visible light. It means that unless a body is very hot, like the Sun, it will not have enough energy to give off even a single quantum of ultra-violet or X-rays. That is why we don't get sunburn from a cup of coffee.

Planck regarded the idea of quanta as just a mathematical trick, and not as having any physical reality, whatever that might mean. However, physicists began to find other behaviour that could be explained only in terms of quantities having discrete or quantised values rather than continuously variable ones. For example, it was found that elementary particles behaved rather like little tops, spinning about an axis. But the amount of spin couldn't have just any value. It had to be some multiple of a basic unit. Because this unit is very small, one does not notice that a normal top really slows down in a rapid sequence of discrete steps, rather than as a continuous process. But, for tops as small as atoms, the discrete nature of spin is very important.

It was some time before people realised the implications of this quantum behaviour for determinism. It was not until 1927 that Werner Heisenberg, another German physicist, pointed out that you couldn't measure simultaneously both the position and speed of a particle exactly. To see where a particle is, one has to shine light on it. But by Planck's work one can't use an arbitrarily small amount of light. One has to use at least one quantum. This will disturb the particle and change its speed in a way that can't be predicted. To measure the position of the particle accurately, you will have to use light of short wavelength, like ultra-violet, X-rays or gamma rays. But again, by Planck's work, quanta of these forms of light have higher energies than those of visible light. So they will disturb the speed of the particle more. It is a no-win situation: the more accurately you try to measure the position of the particle, the less accurately you can know the speed, and vice versa. This is summed up in the Uncertainty Principle that Heisenberg formulated; the uncertainty in the position of a particle times the uncertainty in its speed is always greater than a quantity called Planck's constant, divided by twice the mass of the particle.

Laplace's vision of scientific determinism involved

knowing the positions and speeds of the particles in the universe, at one instant of time. So it was seriously undermined by Heisenberg's Uncertainty Principle. How could one predict the future, when one could not measure accurately both the positions and the speeds of particles at the present time? No matter how powerful a computer you have, if you put lousy data in you will get lousy predictions out.

Einstein was very unhappy about this apparent randomness in nature. His views were summed up in his famous phrase 'God does not play dice'. He seemed to have felt that the uncertainty was only provisional and that there was an underlying reality, in which particles would have well-defined positions and speeds and would evolve according to deterministic laws in the spirit of Laplace. This reality might be known to God, but the quantum nature of light would prevent us seeing it, except through a glass darkly.

Einstein's view was what would now be called a hidden variable theory. Hidden variable theories might seem to be the most obvious way to incorporate the Uncertainty Principle into physics. They form the basis of the mental picture of the universe held by many scientists, and almost all philosophers of science. But these hidden variable theories are wrong. The British physicist John Bell devised an experimental test that could falsify hidden variable theories. When the experiment was carried out carefully, the results were inconsistent with hidden variables. Thus it seems that even God is bound by the Uncertainty Principle and cannot know both the position and the speed of a particle. All the evidence points to God being an inveterate gambler, who throws the dice on every possible occasion.

Other scientists were much more ready than Einstein to modify the classical nineteenth-century view of determinism. A new theory, quantum mechanics, was put forward by Heisenberg, Erwin Schrödinger from Austria and the British physicist Paul Dirac. Dirac was my predecessor but one as the Lucasian Professor in Cambridge. Although quantum mechanics has been around for nearly seventy years, it is still not generally understood or appreciated, even by those who use it to do calculations. Yet it should concern us all, because it is completely different from the classical picture of the physical universe, and of reality itself. In quantum mechanics, particles don't have well-defined positions and speeds. Instead, they are represented by what is called a wave function. This is a number at each point of space. The size of the wave function gives the probability that the particle will be found in that position. The rate at which the wave function varies from point to point gives the speed of the particle. One can have a wave function that is very strongly peaked in a small region. This will mean that the uncertainty in the position is small. But the wave function will vary very rapidly near the peak, up on one side and down on the other. Thus the uncertainty in the speed will be large. Similarly, one can have wave functions where the uncertainty in the speed is small but the uncertainty in the position is large.

The wave function contains all that one can know of the particle, both its position and its speed. If you know the wave function at one time, then its values at other times are determined by what is called the Schrödinger equation. Thus one still has a kind of determinism, but it is not the sort that Laplace envisaged. Instead of being able to predict the positions and speeds of particles, all we can predict is the wave function. This means that we can predict just half what we could according to the classical nineteenth-century view.

Although quantum mechanics leads to uncertainty when we try to predict both the position and the speed, it still allows us to predict, with certainty, one combination of position and speed. However, even this degree of certainty seems to be threatened by more recent developments. The problem arises because gravity can warp space–time so much that there can be regions of space that we can't observe.

Do the laws governing the universe allow us to predict exactly what is going to happen to us in the future?

The short answer is no, and yes. In principle, the laws allow us to predict the future. But in practice the calculations are often too difficult.

Such regions are the interiors of black holes. That means that we cannot, even in principle, observe the particles inside a black hole. So we cannot measure their positions or velocities at all. There is then an issue of whether this introduces further unpredictability beyond that found in quantum mechanics.

To sum up, the classical view, put forward by Laplace, was that the future motion of particles was completely determined, if one knew their positions and speeds at one time. This view had to be modified when Heisenberg put forward his Uncertainty Principle, which said that one could not know both the position and the speed accurately. However, it was still possible to predict one combination of position and speed. But perhaps even this limited predictability might disappear if black holes are taken into account.

5

WHAT IS INSIDE A BLACK HOLE?

It is said that fact is sometimes stranger than fiction, and nowhere is that more true than in the case of black holes. Black holes are stranger than anything dreamed up by science-fiction writers, but they are firmly matters of science fact.

The first discussion of black holes was in 1783, by a Cambridge man, John Michell. His argument ran as follows. If one fires a particle, such as a cannon ball, vertically upwards, it will be slowed down by gravity. Eventually, the particle will stop moving upwards, and will fall back. However, if the initial upwards velocity were greater than some critical value, called the escape velocity, gravity would never be strong enough to stop the particle, and it would get away. The escape velocity is just over 11 kilometres per second for the Earth, and about 617 kilometres per second for the Sun. Both of these are much higher than the speed of real cannon balls. But they are low compared to the speed of light, which is 300,000 kilometres per second. Thus light can get away from the Earth or Sun without much difficulty. However, Michell argued that there could be stars that were much more massive than the Sun which had escape velocities greater than the speed of light. We would not be able to see them, because any light they sent out would be dragged back by gravity. Thus they would be what Michell called dark stars, what we now call black holes.

To understand them, we need to start with gravity. Gravity is described by Einstein's general theory of relativity, which is a theory of space and time as well as gravity. The behaviour of space and time is governed by a set of equations called the Einstein equations which Einstein put forward in 1915. Although gravity is by far the weakest of the known forces of nature, it has two crucial advantages over other forces. First, it acts over a long range. The Earth is held in orbit by the Sun, ninety-three million miles away, and the Sun is held in orbit around the centre of the galaxy, about 10,000 light years away. The second advantage is that gravity is always attractive, unlike electric forces which can be either attractive or repulsive. These two features mean that for a sufficiently large star the gravitational attraction between particles can dominate over all other forces and lead to gravitational collapse. Despite these facts, the scientific community was slow to realise that massive stars could collapse in on themselves under their own gravity and to figure out how the object left behind would behave. Albert Einstein even wrote a paper in 1939 claiming that stars could not collapse under gravity, because matter could not be compressed beyond a certain point. Many scientists shared Einstein's gut feeling. The principal exception was the American scientist John Wheeler, who in many ways is the hero of the black hole story. In his work in the 1950s and 1960s, he emphasised that many stars would eventually collapse, and explored the problems this posed for theoretical physics. He also foresaw many of the properties of the objects which collapsed stars become – that is, black holes.

During most of the life of a normal star, over many billions of years, it will support itself against its own gravity by thermal pressure caused by nuclear processes which convert hydrogen into helium. Eventually, however, the star will exhaust its nuclear fuel. The star will contract. In some cases, it may be able to support itself as a white dwarf star, the dense remnants of a stellar core. However, Subrahmanyan Chandrasekhar showed in 1930 that the maximum mass of a white dwarf star is about 1.4 times that of the Sun. A similar maximum mass was calculated by the Russian physicist Lev Landau for a star made entirely of neutrons.

What would be the fate of those countless stars with a greater mass than the maximum mass of a white dwarf or neutron star once they had exhausted nuclear fuel? The problem was investigated by Robert Oppenheimer of later atom bomb fame. In a couple of papers in 1939, with George Volkoff and Hartland Snyder, he showed that such a star could not be supported by pressure. And that if one neglected pressure, a uniform spherically systematic symmetric star would contract to a single point of infinite density. Such a point is called a singularity. All our theories of space are formulated on the assumption that space–time is smooth and nearly flat, so they break down at the singularity, where the curvature of space–time is infinite. In fact, it marks the end of space and time itself. That is what Einstein found so objectionable.

Then the Second World War intervened. Most scientists, including Robert Oppenheimer, switched their attention to nuclear physics, and the issue of gravitational collapse was largely forgotten. Interest in the subject revived with the discovery of distant objects called quasars. The first quasar, 3C273, was found in 1963. Many other quasars were soon discovered. They were bright despite being at great distances from the Earth. Nuclear processes could not account for their energy output, because they release only a small fraction of their rest mass as pure energy. The only alternative was gravitational energy released by gravitational collapse.

Gravitational collapse of stars was rediscovered. When this happens, the gravity of the object draws all its surrounding matter inwards. It was clear that a uniform spherical star would contract to a point of infinite density, a singularity. But what would happen if the star isn't uniform and spherical? Could this unequal distribution of the star's matter cause a non-uniform collapse and avoid a singularity? In a remarkable paper in 1965, Roger Penrose showed there would still be a singularity, using only the fact that gravity is attractive.

The Einstein equations can't be defined at a singularity. This means that at this point of infinite density one can't predict the future. This implies that something strange could happen whenever a star collapsed. We wouldn't be affected by the breakdown of prediction if the singularities are not naked – that is, they are not shielded from the outside. Penrose proposed the cosmic censorship conjecture: all singularities formed by the collapse of stars or other bodies are hidden from view inside black holes. A black hole is a region where gravity is so strong that light cannot escape. The cosmic censorship conjecture is almost certainly true, because a number of attempts to disprove it have failed.

When John Wheeler introduced the term 'black hole'

in 1967, it replaced the earlier name of 'frozen star'. Wheeler's coinage emphasised that the remnants of collapsed stars are of interest in their own right, independently of how they were formed. The new name caught on quickly.

From the outside, you can't tell what is inside a black hole. Whatever you throw in, or however it is formed, black holes look the same. John Wheeler is known for expressing this principle as 'A black hole has no hair.'

A black hole has a boundary called the event horizon. It is where gravity is just strong enough to drag light back and prevent it from escaping. Because nothing can travel faster than light, everything else will get dragged back also. Falling through the event horizon is a bit like going over Niagara Falls in a canoe. If you are above the Falls, you can get away if you paddle fast enough, but once you are over the edge you are lost. There's no way back. As you get nearer the Falls, the current gets faster. This means it pulls harder on the front of the canoe than the back. There's a danger that the canoe will be pulled apart. It is the same with black holes. If you fall towards a black hole feet first, gravity will pull harder on your feet than your head, because they are nearer the black hole. The result is that you will be stretched out lengthwise, and squashed in sideways. If the black hole has a mass of a few times our Sun, you would be torn apart and made into spaghetti before you reached the horizon. However, if you fell into a much larger black hole, with a mass of more than a million times the Sun, the gravitational pull would be the same on the whole of your body and you would reach the horizon without difficulty. So, if you want to explore the inside of a black hole, make sure you choose a big one. There is a black hole with a mass of about four million times that of the Sun at the centre of our Milky Way galaxy.

Although you wouldn't notice anything in particular as you fell into a black hole, someone watching you from a distance would never see you cross the event horizon. Instead, you would appear to slow down and hover just outside. Your image would get dimmer and dimmer, and redder and redder, until you were effectively lost from sight. As far as the outside world is concerned, you would be lost for ever.

Shortly after the birth of my daughter Lucy I had a eureka moment. I discovered the area theorem. If general relativity is correct, and the energy density of matter is positive, as is usually the case, then the surface area of the event horizon, the boundary of a black hole, has the property that it always increases when additional matter or radiation falls into the black hole. Moreover, if two black holes collide and merge to form a single black hole, the area of the event horizon around the resulting black hole is greater than the sum of the areas of the event horizons around the original black holes. The area theorem can be tested experimentally by the Laser Interferometer Gravitational-Wave Observatory (LIGO). On 14 September 2015, LIGO detected gravitational waves from the collision and merger of two black holes. From the waveform, one can estimate the masses and angular momenta of the black holes, and by the no-hair theorem these determine the horizon areas.

These properties suggest that there is a resemblance between the area of the event horizon of a black hole and conventional classical physics, specifically the concept of entropy in thermodynamics. Entropy can be regarded as a measure of the disorder of a system, or equivalently as a lack of knowledge of its precise state. The famous second law of thermodynamics says that entropy always increases with time. This discovery was the first hint of this crucial connection.

The analogy between the properties of black holes and the laws of thermodynamics can be extended. The first law of thermodynamics says that a small change in the entropy of a system is accompanied by a proportional change in the energy of the system. Brandon Carter, Jim Bardeen and I found a similar law relating the change in mass of a black hole to a change in the area of the event horizon. Here the factor of proportionality involves a quantity called the surface gravity, which is a measure of the strength of the gravitational field at the event horizon. If one accepts that the area of the event horizon is analogous to entropy, then it would seem that the surface gravity is analogous to temperature. The resemblance is strengthened by the fact that the surface gravity turns out to be the same at all points on the event horizon, just as the temperature is the same everywhere in a body at thermal equilibrium.

Although there is clearly a similarity between entropy and the area of the event horizon, it was not obvious to us how the area could be identified as the entropy of a black hole itself. What would be meant by the entropy of a black hole? The crucial suggestion was made in 1972 by Jacob Bekenstein, who was a graduate student at Princeton University. It goes like this. When a black hole is created by gravitational collapse, it rapidly settles down to a stationary state, which is characterised by three parameters: the mass, the angular momentum and the electric charge.

This makes it look as if the final black hole state is independent of whether the body that collapsed was composed of matter or antimatter, or whether it was spherical or highly irregular in shape. In other words, a black hole of a given mass, angular momentum and electric charge could have been formed by the collapse of any one of a large number of different configurations of matter. So what appears to be the same black hole could be formed by the collapse of a large number of different types of star. Indeed, if quantum effects are neglected, the number of configurations would be infinite since the black hole could have been formed by the collapse of a cloud of an indefinitely large number of particles of indefinitely low mass. But could the number of configurations really be infinite?

Quantum mechanics famously involves the Uncertainty Principle. This states that it is impossible to measure both the position and speed of any object. If one measures exactly where something is, then its speed is undetermined. If one measures the speed of something, then its position is undetermined. In practice, this means that it is impossible to localise anything. Suppose you want to measure the size of something, then you need to figure out where the ends of this moving object are. You can never do this accurately, because it will involve making a measurement of both the positions of something and its speed at the same time. In turn, it is then impossible to determine the size of an object. All you can do is to say that the Uncertainty Principle makes it impossible to say precisely what the size of something really is. It turns out that the Uncertainty Principle imposes a limit on the size of something. After a little bit of calculation, one finds that for a given mass of an object, there is a minimum size. This minimum size is small for heavy objects, but as one looks at lighter and lighter objects, the minimum size gets bigger and bigger. This minimum size can be thought of as a consequence of the fact that in quantum mechanics objects can be thought of either as a wave or a particle. The lighter an object is, the longer its wavelength is and so it is more spread out. The heavier an object is, the shorter its wavelength and so it will seem more compact. When these ideas are combined with those of general relativity, it means that only objects heavier than a particular weight can form black holes. That weight is about the same as that of a grain of salt. A further consequence of these ideas is that the number of configurations that could form a black hole of a given mass, angular momentum, and electric charge, although very large, may also be finite. Jacob Bekenstein suggested that from this finite number, one could interpret the entropy of a black hole. This would be a measure of the amount of information that seems irretrievably lost, during the collapse when a black hole is created.

The apparently fatal flaw in Bekenstein's suggestion was that, if a black hole has a finite entropy that is proportional to the area of its event horizon, it also ought to have a non-zero temperature which would be proportional to its surface gravity. This would imply that a black hole could be in equilibrium with thermal radiation at some temperature other than zero. Yet according to classical concepts no such equilibrium is possible since the black hole would absorb any thermal radiation that fell on it but by definition would not be able to emit anything in return. It cannot emit anything, it cannot emit heat.

This created a paradox about the nature of black holes, the incredibly dense objects created by the collapse of stars. One theory suggested that black holes with identical qualities could be formed from an infinite number of different types of stars. Another suggested that the number could be finite. This is a problem of information – the idea that every particle and every force in the universe contains information.

Because black holes have no hair, as the scientist John Wheeler put it, one can't tell from the outside what is inside a black hole, apart from its mass, electric charge and rotation. This means that a black hole must contain a lot of information that is hidden from the outside world. But there is a limit to the amount of information one can pack into a region of space. Information requires energy, and energy has mass by Einstein's famous equation, $E = mc^2$. So, if there's too much information in a region of space, it will collapse into a black hole, and the size of the black hole will reflect the amount of information. It is like piling more and more books into a library. Eventually, the shelves will give way and the library will collapse into a black hole.

If the amount of hidden information inside a black hole depends on the size of the hole, one would expect from general principles that the black hole would have a temperature and would glow like a piece of hot metal. But that was impossible because, as everyone knew, nothing could get out of a black hole. Or so it was thought.

This problem remained until early in 1974, when I was investigating what the behaviour of matter in the vicinity of a black hole would be according to quantum mechanics. To my great surprise, I found that the black hole seemed to emit particles at a steady rate. Like everyone else at that time, I accepted the dictum that a black hole could not emit anything. I therefore put quite a lot of effort into trying to get rid of this embarrassing effect. But the more I thought about it, the more it refused to go away, so that in the end I had to accept it. What finally convinced me it was a real physical process was that the outgoing particles have a spectrum that is precisely thermal. My calculations predicted that a black hole creates and emits particles and radiation, just as if it were an ordinary hot body, with a temperature that is proportional to the surface gravity and inversely proportional to the mass. This made the problematic suggestion of Jacob Bekenstein, that a black

Is falling into a black hole bad news for a space traveller?

Definitely bad news. If it were a stellar mass black hole, you would be made into spaghetti before reaching the horizon. On the other hand, if it were a supermassive black hole, you would cross the horizon with ease, but be crushed out of existence at the singularity.

hole had a finite entropy, fully consistent, since it implied that a black hole could be in thermal equilibrium at some finite temperature other than zero.

Since that time, the mathematical evidence that black holes emit thermal radiation has been confirmed by a number of other people with various different approaches. One way to understand the emission is as follows. Quantum mechanics implies that the whole of space is filled with pairs of virtual particles and anti-particles that are constantly materialising in pairs, separating and then coming together again, and annihilating each other. These particles are called virtual, because, unlike real particles, they cannot be observed directly with a particle detector. Their indirect effects can nonetheless be measured, and their existence has been confirmed by a small shift, called the Lamb shift, which they produce in the spectrum energy of light from excited hydrogen atoms. Now, in the presence of a black hole, one member of a pair of virtual particles may fall into the hole, leaving the other member without a partner with which to engage in mutual annihilation. The forsaken particle or antiparticle may fall into the black hole after its partner, but it may also escape to infinity, where it appears to be radiation emitted by the black hole.

Another way of looking at the process is to regard the member of the pair of particles that falls into the black hole, the antiparticle say, as being really a particle that is travelling backwards in time. Thus the antiparticle falling into the black hole can be regarded as a particle coming out of the black hole but travelling backwards in time. When the particle reaches the point at which the particle–antiparticle pair originally materialised, it is scattered by the gravitational field, so that it travels forward in time. A black hole of the mass of the Sun would leak particles at such a slow rate that it would be impossible to detect. However, there could be much smaller mini black holes with the mass of, say, a mountain. These might have formed in the very early universe if it had been chaotic and irregular. A mountain-sized black hole would give off X-rays and gamma rays, at a rate of about ten million megawatts, enough to power the world's electricity supply. It wouldn't be easy, however, to harness a mini black hole. You couldn't keep it in a power station because it would drop through the floor and end up at the centre of the Earth. If we had such a black hole, about the only way to keep hold of it would be to have it in orbit around the Earth.

People have searched for mini black holes of this mass, but have so far not found any. This is a pity because, if they had, I would have got a Nobel Prize. Another possibility, however, is that we might be able to create micro black holes in the extra dimensions of space–time. According to some theories, the universe we experience is just a four-dimensional surface in a ten- or eleven-dimensional space. The movie *Interstellar* gives some idea of what this is like. We wouldn't see these extra dimensions, because light wouldn't propagate through them but only through the four dimensions of our universe. Gravity however, would affect the extra dimensions, and would be much stronger than in our universe. This would make it much easier to form a little black hole in the extra dimensions. It might be possible to observe this at the LHC, the Large Hadron Collider, at CERN in Switzerland. This consists of a circular tunnel, twenty-seven kilometres long. Two beams of particles travel round this tunnel in opposite directions and are made to collide. Some of the collisions might create micro black holes. These would radiate particles in a pattern that would be easy to recognise. So I might get a Nobel Prize after all.*

As particles escape from a black hole, the hole will lose mass and shrink. This will increase the rate of emission of particles. Eventually, the black hole will lose all its mass and disappear. What then happens to all the particles and unlucky astronauts that fell into the black hole? They can't just re-emerge when the

---

* Nobel Prizes cannot be awarded posthumously, so sadly this ambition will never be realised.

black hole disappears. The particles that come out of a black hole seem to be completely random and to bear no relation to what fell in. It appears that the information about what fell in is lost, apart from the total amount of mass and the amount of rotation. But if information is lost, this raises a serious problem that strikes at the heart of our understanding of science. For more than 200 years, we have believed in scientific determinism – that is, that the laws of science determine the evolution of the universe.

If information were really lost in black holes, we wouldn't be able to predict the future, because a black hole could emit any collection of particles. It could emit a working television set or a leather-bound volume of the complete works of Shakespeare, though the chance of such exotic emissions is very low. It is much more likely to emit thermal radiation, like the glow from red-hot metal. It might seem that it wouldn't matter very much if we couldn't predict what comes out of black holes. There aren't any black holes near us. But it is a matter of principle. If determinism, the predictability of the universe, breaks down with black holes, it could break down in other situations. There could be virtual black holes that appear as fluctuations out of the vacuum, absorb one set of particles, emit another and disappear into the vacuum again. Even worse, if determinism breaks down, we can't be sure of our past history either. The history books and our memories could just be illusions. It is the past that tells us who we are. Without it, we lose our identity.

It was therefore very important to determine whether information really was lost in black holes, or whether in principle it could be recovered. Many scientists felt that information should not be lost, but for years no one suggested a mechanism by which it could be preserved. This apparent loss of information, known as the information paradox, has troubled scientists for the last forty years, and still remains one of the biggest unsolved problems in theoretical physics.

Recently, interest in possible resolutions of the information paradox has been revived as new discoveries have been made about the unification of gravity and quantum mechanics. Central to these recent break-throughs is the understanding of the symmetries of space–time.

Suppose there was no gravity and space–time was completely flat. This would be like a completely feature-less desert. Such a place has two types of symmetry. The first is called translation symmetry. If you moved from one point in the desert to another, you would not notice any change. The second symmetry is rotation symmetry. If you stood somewhere in the desert and started to turn around, you would again not notice any difference in what you saw. These symmetries are also found in 'flat' space–time, the space–time one finds in the absence of any matter.

If one put something into this desert, these symmetries would be broken. Suppose there was a mountain, an oasis and some cacti in the desert, it would look different in different places and in different directions. The same is true of space–time. If one puts objects into a space–time, the translational and rotational symmetries get broken. And introducing objects into a space–time is what produces gravity.

A black hole is a region of space–time where gravity is strong, space–time is violently distorted and so one expects its symmetries to be broken. However, as one moves away from the black hole, the curvature of space–time gets less and less. Very far away from the black hole, space–time looks very much like flat space–time.

Back in the 1960s, Hermann Bondi, A. W. Kenneth Metzner, M. G. J. van der Burg and Rainer Sachs made the truly remarkable discovery that space–time far away from any matter has an infinite collection of symmetries known as supertranslations. Each of these symmetries is associated with a conserved quantity, known as the supertranslation charges. A conserved quantity is a quantity that does not change as a system evolves. These are generalisations of more familiar conserved quantities. For example, if space–time does not change in time, then energy is conserved. If space–time looks the same at different points in space, then momentum is conserved.

What was remarkable about the discovery of super-translations is that there are an infinite number of conserved quantities far from a black hole. It is these conservation laws that have given an extraordinary and unexpected insight into process in gravitational physics.

In 2016, together with my collaborators Malcolm Perry and Andy Strominger, I was working on using these new results with their associated conserved quantities to find a possible resolution to the information paradox. We know that the three discernible properties of black holes are their mass, their charge and their angular momentum. These are the classical charges that have been understood for a long time. However, black holes also carry a supertranslation charge. So perhaps black holes have a lot more to them than we first thought. They are not bald or with only three hairs, but actually have a very large amount of supertranslation hair.

This supertranslation hair might encode some of the information about what is inside the black hole. It is likely that these supertranslation charges do not contain all of the information, but the rest might be accounted for by some additional conserved quantities, super-rotation charges, associated with some additional related symmetries called superrotations, which arc, as yet, not well understood. If this is right, and all the information about a black hole can be understood in terms of its 'hairs', then perhaps there is no loss of information. These ideas have just received confirmation with our most recent calculations. Strominger, Perry and myself, together with a graduate student, Sasha Haco, have discovered that these superrotation charges can account for the entire entropy of any black hole. Quantum mechanics continues to hold, and information is stored on the horizon, the surface of the black hole.

The black holes are still characterised only by their overall mass, electric charge and spin outside the event horizon but the event horizon itself contains the information needed to tell us about what has fallen into the black hole in a way that goes beyond these three characteristics the black hole has. People are still working on these issues and therefore the information paradox remains unresolved. But I am optimistic that we are moving towards a solution. Watch this space.

6

IS TIME TRAVEL POSSIBLE?

In science fiction, space and time warps are common-place. They are used for rapid journeys around the galaxy or for travel through time. But today's science fiction is often tomorrow's science fact. So what are the chances of time travel?

The idea that space and time can be curved or warped is fairly recent. For more than 2,000 years the axioms of Euclidean geometry were considered to be self-evident. As those of you who were forced to learn geometry at school may remember, one of the consequences of these axioms is that the angles of a triangle add up to 180 degrees.

However, in the last century people began to realise that other forms of geometry were possible in which the angles of a triangle need not add up to 180 degrees. Consider, for example, the surface of the Earth. The nearest thing to a straight line on the surface of the Earth is what is called a great circle. These are the shortest paths between two points so they are the routes that airlines use. Consider now the triangle on the surface of the Earth made up of the equator, the line of 0 degrees longitude through London and the line of 90 degrees longtitude east through Bangladesh. The two lines of longitude meet the equator at a right angle, or 90 degrees. The two lines of longitude also meet each other at the North Pole at a right angle, or 90 degrees. Thus one has a triangle with three right angles. The angles of this triangle add up to 270 degrees which is obviously greater than the 180 degrees for a triangle on a flat surface. If one drew a triangle on a saddle-shaped surface one would find that the angles added up to less than 180 degrees.

The surface of the Earth is what is called a two-dimensional space. That is, you can move on the surface of the Earth in two directions at right angles to each other: you can move north–south or east–west. But of course there is a third direction at right angles to these two and that is up or down. In other words the surface of the Earth exists in three-dimensional space. The three-dimensional space is flat. That is to say it obeys Euclidean geometry. The angles of a triangle add up to 180 degrees. However, one could imagine a race of two-dimensional creatures who could move about on the surface of the Earth but who couldn't experience the third direction of up or down. They wouldn't know about the flat three-dimensional space in which the surface of the Earth lives. For them space would be curved and geometry would be non-Euclidean.

But just as one can think of two-dimensional beings living on the surface of the Earth, so one could imagine that the three-dimensional space in which we live was the surface of a sphere in another dimension that we don't see. If the sphere were very large, space would be nearly flat and Euclidean geometry would be a very good approximation over small distances. But we would notice that Euclidean geometry broke down over large distances. As an illustration of this imagine a team of painters adding paint to the surface of a large ball.

As the thickness of the paint layer increased, the surface area would go up. If the ball were in a flat three-dimensional space one could go on adding paint indefinitely and the ball would get bigger and bigger. However, if the three-dimensional space were really the surface of a sphere in another dimension its volume would be large but finite. As one added more layers of paint the ball would eventually fill half the space. After that the painters would find that they were trapped in a region of ever-decreasing size, and almost the whole of space would be occupied by the ball and its layers of paint. So they would know that they were living in a curved space and not a flat one.

This example shows that one cannot deduce the geometry of the world from first principles as the ancient Greeks thought. Instead one has to measure the space we live in and find out its geometry by experiment. However, although a way to describe curved spaces was developed by the German Bernhard Riemann in 1854, it remained just a piece of mathematics for sixty years. It could describe curved spaces that existed in the abstract, but there seemed no reason why the physical space we lived in should be curved. This reason came only in 1915 when Einstein put forward the general theory of relativity.

General relativity was a major intellectual revolution that has transformed the way we think about the universe. It is a theory not only of curved space but of curved or warped time as well. Einstein had realised in 1905 that space and time are intimately connected with each other, which is when his theory of special relativity was born, relating space and time to each other. One can describe the location of an event by four numbers. Three numbers describe the position of the event. They could be miles north and east of Oxford Circus and the height above sea level. On a larger scale they could be galactic latitude and longitude and distance from the centre of the galaxy.

The fourth number is the time of the event. Thus one can think of space and time together as a four-dimensional entity called space–time. Each point of space–time is labelled by four numbers that specify its position in space and in time. Combining space and time into space–time in this way would be rather trivial if one could disentangle them in a unique way. That is to say if there was a unique way of defining the time and position of each event. However, in a remarkable paper written in 1905 when he was a clerk in the Swiss patent office, Einstein showed that the time and position at which one thought an event occurred depended on how one was moving. This meant that time and space were inextricably bound up with each other.

The times that different observers would assign to events would agree if the observers were not moving relative to each other. But they would disagree more the faster their relative speed. So one can ask how fast does one need to go in order that the time for one observer should go backwards relative to the time of another observer. The answer is given in the following limerick:

There was a young lady of Wight
Who travelled much faster than light
She departed one day
In a relative way
And arrived on the previous night.

So all we need for time travel is a spaceship that will go faster than light. Unfortunately in the same paper Einstein showed that the rocket power needed to accelerate a spaceship got greater and greater the nearer it got to the speed of light. So it would take an infinite amount of power to accelerate past the speed of light.

Einstein's paper of 1905 seemed to rule out time travel into the past. It also indicated that space travel to other stars was going to be a very slow and tedious business. If one couldn't go faster than light the round trip from us to the nearest star would take at least eight years and to the centre of the galaxy about 50,000 years. If the spaceship went very near the speed of light it might seem to the people on board that the trip to the galactic centre had taken only a few years. But that wouldn't be much consolation if everyone you had known had died and been forgotten thousands of years ago when you got back. That wouldn't be much good for science-fiction novels either, so writers had to look for ways to get round this difficulty.

In 1915, Einstein showed that the effects of gravity could be described by supposing that space–time was warped or distorted by the matter and energy in it, and this theory is known as general relativity. We can actually observe this warping of space–time produced by the mass of the Sun in the slight bending of light or radio waves passing close to the Sun.

This causes the apparent position of the star or radio source to shift slightly when the Sun is between the Earth and the source. The shift is very small, about a thousandth of a degree, equivalent to a movement of an inch at a distance of a mile. Nevertheless it can be measured with great accuracy and it agrees with the predictions of general relativity. We have experimental evidence that space and time are warped.

The amount of warping in our neighbourhood is very small because all the gravitational fields in the solar system are weak. However, we know that very strong fields can occur, for example in the Big Bang or in black holes. So can space and time be warped enough to meet the demands from science fiction for things like hyperspace drives, wormholes or time travel? At first sight all these seem possible. For example, in 1948 Kurt Gödel found a solution to Einstein's field equations of general relativity that represents a universe in which all the matter was rotating. In this universe it would be possible to go off in a spaceship and come back before you had set out. Gödel was at the Institute of Advanced Study in Princeton, where Einstein also spent his last years. He was more famous for proving you couldn't prove everything that is true even in such an apparently simple subject as arithmetic. But what he proved about general relativity allowing time travel really upset Einstein, who had thought it wouldn't be possible.

We now know that Gödel's solution couldn't represent the universe in which we live because it was not expanding. It also had a fairly large value for a quantity called the cosmological constant which is generally believed to be very small. However, other apparently more reasonable solutions that allow time travel have since been found. A particularly interesting one from an approach known as string theory contains two cosmic strings moving past each other at a speed very near to but slightly less than the speed of light. Cosmic strings are a remarkable idea of theoretical physics which science-fiction writers don't really seem to have caught on to. As their name suggests they are like string in that they have length but a tiny cross-section. Actually they are more like rubber bands because they are under enormous tension, something like a hundred billion billion billion tonnes. A cosmic string attached to the Sun would accelerate it from nought to sixty in a thirtieth of a second.

Cosmic strings may sound far-fetched and pure science fiction, but there are good scientific reasons to believe they could have formed in the very early universe shortly after the Big Bang. Because they are under such great tension one might have expected them to accelerate to almost the speed of light.

What both the Gödel universe and the fast-moving cosmic-string space–time have in common is that they start out so distorted and curved that space–time curves back on itself and travel into the past was always possible. God might have created such a warped universe, but we have no reason to think that he did. All the evidence is that the universe started out in the Big Bang without the kind of warping needed to allow travel into the past. Since we can't change the way the universe began, the question of whether time travel is possible is one of whether we can subsequently make space–time so warped that one can go back to the past. I think this is an important subject for research, but one has to be careful not to be labelled a crank. If one made a research grant application to work on time travel it would be dismissed immediately. No government agency could afford to be seen to be spending public money on anything as way out as time travel. Instead one has to use technical terms like closed time-like curves, which are code for time travel. Yet it is a very serious question. Since general relativity can permit time travel, does it allow it in our universe? And if not, why not?

Closely related to time travel is the ability to travel rapidly from one position in space to another. As I said earlier, Einstein showed that it would take an infinite amount of rocket power to accelerate a spaceship to beyond the speed of light. So the only way to get from one side of the galaxy to the other in a reasonable time would seem to be if we could warp space–time so much that we created a little tube or wormhole. This could connect the two sides of the galaxy and act as a short cut to get from one to the other and back while your friends were still alive. Such wormholes have been seriously suggested as being within the capabilities of a future civilisation. But if you can travel from one side of the galaxy to the other in a week or two you could go back through another wormhole and arrive back before you had set out. You could even manage to travel back in time with a single wormhole if its two ends were moving relative to each other.

One can show that to create a wormhole one needs to warp space–time in the opposite way to that in which normal matter warps it. Ordinary matter curves space–time back on itself, like the surface of the Earth. However, to create a wormhole one needs matter that warps space–time in the opposite way, like the surface of a saddle. The same is true of any other way of warping space–time to allow travel to the past if the universe didn't begin so warped that it allowed time travel. What one would need would be matter with negative mass and negative energy density to make space–time warp in the way required.

Energy is rather like money. If you have a positive bank balance, you can distribute it in various ways. But, according to the classical laws that were believed until quite recently, you weren't allowed to have an energy overdraft. So these classical laws would have ruled out us being able to warp the universe in the way required to allow time travel. However, the classical laws were overthrown by quantum theory, which is the other great revolution in our picture of the universe apart from general relativity. Quantum theory is more relaxed and allows you to have an overdraft on one or two accounts. If only the banks were as accommodating. In other words, quantum theory allows the energy density to be negative in some places provided it is positive in others.

The reason quantum theory can allow the energy density to be negative is that it is based on the Uncertainty Principle. This says that certain quantities like the position and speed of a particle can't both have well-defined values. The more accurately the position of a particle is defined the greater is the uncertainty in its speed, and vice versa. The Uncertainty Principle also applies to fields like the electromagnetic field or the gravitational field. It implies that these fields can't be exactly zero even in what we think of as empty space. For if they were exactly zero their values would have both a well-defined position at zero and a well-defined speed which was also zero. This would be a violation of the Uncertainty Principle. Instead the fields would have to have a certain minimum amount of fluctuations.

One can interpret these so-called vacuum fluctuations as pairs of particles and antiparticles that suddenly appear together, move apart and then come back together again and annihilate each other.

These particle–antiparticle pairs are said to be virtual because one cannot measure them directly with a particle detector. However, one can observe their effects indirectly. One way of doing this is by what is called the Casimir effect. Imagine that you have two parallel metal plates a short distance apart. The plates act like mirrors for the virtual particles and anti-particles. This means that the region between the plates is a bit like an organ pipe and will only admit light waves of certain resonant frequencies. The result is that there are a slightly different number of vacuum fluctuations or virtual particles between the plates than there are outside them, where vacuum fluctuations can have any wavelength. The difference in the number of virtual particles between the plates compared with outside the plates means that they don't exert as much pressure on one side of the plates compared with the other. There is thus a slight force pushing the plates together. This force has been measured experimentally. So, virtual particles actually exist and produce real effects.

Because there are fewer virtual particles or vacuum fluctuations between the plates, they have a lower energy density than in the region outside. But the energy density of empty space far away from the plates must be zero. Otherwise it would warp space–time and the universe wouldn't be nearly flat. So the energy density in the region between the plates must be negative.

We thus have experimental evidence from the bending of light that space–time is curved and confirmation from the Casimir effect that we can warp it in the negative direction. So it might seem that as we advance in science and technology we might be able to construct a wormhole or warp space and time in some other way so as to be able to travel into our past. If this were the case it would raise a whole host of questions and problems. One of these is if time travel will be possible in the future, why hasn't someone come back from the future to tell us how to do it.

Even if there were sound reasons for keeping us in ignorance, human nature being what it is it is difficult to believe that someone wouldn't show off and tell us poor benighted peasants the secret of time travel. Of course, some people would claim that we have already been visited from the future. They would say that UFOs come from the future and that governments are engaged in a gigantic conspiracy to cover them up and keep for themselves the scientific knowledge that these visitors bring. All I can say is that if governments were hiding something they are doing a poor job of extracting useful information from the aliens. I'm pretty sceptical of conspiracy theories, as I believe that cock-up theory is more likely. The reports of sightings of UFOs cannot all be caused by extra-terrestrials because they are mutually contradictory. But, once you admit that some are mistakes or hallucinations, isn't it more probable that they all are than that we are being visited by people from the future or from the other side of the galaxy? If they really want to colonise the Earth or warn us of some danger they are being rather ineffective.

A possible way to reconcile time travel with the fact that we don't seem to have had any visitors from the future would be to say that such travel can occur only in the future. In this view one would say space–time in our past was fixed because we have observed it and seen that it is not warped enough to allow travel into the past. On the other hand the future is open. So we might be able to warp it enough to allow time travel. But because we can warp space–time only in the future, we wouldn't be able to travel back to the present time or earlier.

This picture would explain why we haven't been overrun by tourists from the future. But it would still leave plenty of paradoxes. Suppose it were possible to go off in a rocket ship and come back before you had set off. What would stop you blowing up the rocket on its launch pad or otherwise preventing yourself from setting out in the first place? There are other versions of this paradox, like going back and killing your parents before you were born, but they are essentially equivalent. There seem to be two possible resolutions.

One is what I shall call the consistent-histories approach. It says that one has to find a consistent solution of the equations of physics even if space–time is so warped that it is possible to travel into the past. On this view you couldn't set out on the rocket ship to travel into the past unless you had already come back and failed to blow up the launch pad. It is a consistent picture, but it would imply that we were completely determined: we couldn't change our minds. So much for free will.

The other possibility is what I call the alternative-histories approach. It has been championed by the physicist David Deutsch and it seems to have been what the creator of *Back to the Future* had in mind. In this view, in one alternative history there would not have been any return from the future before the rocket set off and so no possibility of it being blown up. But when the traveller returns from the future he enters another alternative history. In this the human race makes a tremendous effort to build a spaceship, but just before it is due to be launched a similar spaceship appears from the other side of the galaxy and destroys it.

David Deutsch claims support for the alternative-histories approach from the sum-over-histories concept introduced by the physicist Richard Feynman. The idea is that according to quantum theory the universe doesn't just have a unique single history. Instead the universe has every single possible history, each with its own probability. There must be a possible history in which there is a lasting peace in the Middle East, though maybe the probability is low.

In some histories space–time will be so warped that objects like rockets will be able to travel into their pasts. But each history is complete and self-contained, describing not only the curved space–time but also the objects in it. So a rocket cannot transfer to another alternative history when it comes round again. It is still in the same history which has to be self-consistent. Thus despite what Deutsch claims I think the sum-over-histories idea supports the consistent-histories hypothesis rather than the alternative-histories idea.

It thus seems that we are stuck with the consistent-histories picture. However, this need not involve problems with determinism or free will if the probabilities are very small for histories in which space–time is so warped that time travel is possible over a macroscopic region. This is what I call the Chronology Protection Conjecture: the laws of physics conspire to prevent time travel on a macroscopic scale.

It seems that what happens is that when space–time gets warped almost enough to allow travel into the past

Is there any point in hosting a party for time travellers? Would you hope anyone would turn up?

In 2009 I held a party for time travellers in my college, Gonville and Caius in Cambridge, for a film about time travel. To ensure that only genuine time travellers came, I didn't send out the invitations until after the party. On the day of the party, I sat in college hoping, but no one came. I was disappointed, but not surprised, because I had shown that if general relativity is correct and energy density is positive, time travel is not possible. I would have been delighted if one of my assumptions had turned out to be wrong.

virtual particles can almost become real particles following closed trajectories. The density of the virtual particles and their energy become very large. This means that the probability of these histories is very low. Thus it seems there may be a Chronology Protection Agency at work making the world safe for historians. But this subject of space and time warps is still in its infancy. According to a unifying form of string theory known as M-theory, which is our best hope of uniting general relativity and quantum theory, space–time ought to have eleven dimensions, not just the four that we experience. The idea is that seven of these eleven dimensions are curled up into a space so small that we don't notice them. On the other hand the remaining four directions are fairly flat and are what we call space–time. If this picture is correct it might be possible to arrange that the four flat directions get mixed up with the seven highly curved or warped directions. What this would give rise to we don't yet know. But it opens exciting possibilities.

In conclusion, rapid space travel and travel back in time can't be ruled out according to our present understanding. They would cause great logical problems, so let's hope there's a Chronology Protection Law to prevent people going back and killing their parents. But science-fiction fans need not lose heart. There's hope in M-theory.

7

WILL WE SURVIVE
ON EARTH?

In January 2018, the *Bulletin of the Atomic Scientists*, a journal founded by some of the physicists who had worked on the Manhattan Project to produce the first atomic weapons, moved the Doomsday Clock, their measurement of the imminence of catastrophe – military or environmental – facing our planet, forward to two minutes to midnight.

The clock has an interesting history. It was started in 1947, at a time when the atomic age had only just begun. Robert Oppenheimer, the chief scientist for the Manhattan Project, said later of the first explosion of an atomic bomb two years earlier in July 1945, 'We knew the world would not be the same. A few people laughed, a few people cried, most people were silent. I remembered the line from the Hindu scripture, the Bhagavad-Gita, "Now, I am become Death, the destroyer of worlds."'

In 1947, the clock was originally set at seven minutes to midnight. It is now closer to Doomsday than at any time since then, save in the early 1950s at the start of the Cold War. The clock and its movements are, of course, entirely symbolic but I feel compelled to point out that such an alarming warning from other scientists, prompted at least in part by the election of Donald Trump, must be taken seriously. Is the clock, and the idea that time is ticking or even running out for the human race, realistic or alarmist? Is its warning timely or time-wasting?

I have a very personal interest in time. Firstly, my bestselling book, and the main reason that I am known beyond the confines of the scientific community, was called *A Brief History of Time*. So some might imagine that I am an expert on time, although of course these days an expert is not necessarily a good thing to be. Secondly, as someone who at the age of twenty-one was told by their doctors that they had only five years to live, and who turned seventy-six in 2018, I am an expert on time in another sense, a much more personal one. I am uncomfortably, acutely aware of the passage of time, and have lived much of my life with a sense that the time that I have been granted is, as they say, borrowed.

It is without doubt the case that our world is more politically unstable than at any time in my memory.

Large numbers of people feel left behind both economically and socially. As a result, they are turning to populist – or at least popular – politicians who have limited experience of government and whose ability to take calm decisions in a crisis has yet to be tested. So that would imply that a Doomsday Clock should be moved closer to a critical point, as the prospect of careless or malicious forces precipitating Armageddon grows.

The Earth is under threat from so many areas that it is difficult for me to be positive. The threats are too big and too numerous.

First, the Earth is becoming too small for us. Our physical resources are being drained at an alarming rate. We have presented our planet with the disastrous gift of climate change. Rising temperatures, reduction of the polar ice caps, deforestation, over-population, disease, war, famine, lack of water and decimation of animal species; these are all solvable but so far have not been solved.

Global warming is caused by all of us. We want cars, travel and a better standard of living. The trouble is, by the time people realise what is happening, it may be too late. As we stand on the brink of a Second Nuclear Age and a period of unprecedented climate change, scientists have a special responsibility, once again, to inform the public and to advise leaders about the perils that humanity faces. As scientists, we understand the dangers of nuclear weapons, and their devastating effects, and we are learning how human activities and technologies are affecting climate systems in ways that may forever change life on Earth. As citizens of the world, we have a duty to share that knowledge, and to alert the public to the unnecessary risks that we live with every day. We foresee great peril if governments and societies do not take action now, to render nuclear weapons obsolete and to prevent further climate change.

At the same time, many of those same politicians are denying the reality of man-made climate change, or at least the ability of man to reverse it, just at the moment that our world is facing a series of critical environmental crises. The danger is that global warming may become self-sustaining, if it has not become so already. The melting of the Arctic and Antarctic ice caps reduces the fraction of solar energy reflected back into space, and so increases the temperature further. Climate change may kill off the Amazon and other rainforests and so eliminate one of the main ways in which carbon dioxide is removed from the atmosphere. The rise in sea temperature may trigger the release of large quantities of carbon dioxide. Both these phenomena would increase the greenhouse effect, and so exacerbate global warming. Both effects could make our climate like that of Venus: boiling hot and raining sulphuric acid, but with a temperature of 250 degrees Celsius. Human life would be unsustainable. We need to go beyond the Kyoto Protocol, the international agreement adopted in 1997, and cut carbon emissions now. We have the technology. We just need the political will.

We can be an ignorant, unthinking lot. When we have reached similar crises in our history, there has usually been somewhere else to colonise. Columbus did it in 1492 when he discovered the New World. But now there is no new world. No Utopia around the corner. We are running out of space and the only places to go to are other worlds.

The universe is a violent place. Stars engulf planets, supernovae fire lethal rays across space, black holes bump into each other and asteroids hurtle around at hundreds of miles a second. Granted, these phenomena do not make space sound very inviting, but these are the very reasons why we should venture into space instead of staying put. An asteroid collision would be something against which we have no defence. The last big such collision with us was about sixty-six million years ago and that is thought to have killed the dinosaurs, and it will happen again. This is not science fiction; it is guaranteed by the laws of physics and probability.

Nuclear war is still probably the greatest threat to

humanity at the present time. It is a danger we have rather forgotten. Russia and the United States are no longer so trigger-happy, but suppose there's an accident, or terrorists get hold of the weapons these countries still have. And the risk increases the more countries obtain nuclear weapons. Even after the end of the Cold War, there are still enough nuclear weapons stockpiled to kill us all, several times over, and new nuclear nations will add to the instability. With time, the nuclear threat may decrease, but other threats will develop, so we must remain on our guard.

One way or another, I regard it as almost inevitable that either a nuclear confrontation or environmental catastrophe will cripple the Earth at some point in the next 1,000 years which, as geological time goes, is the mere blink of an eye. By then I hope and believe that our ingenious race will have found a way to slip the surly bonds of Earth and will therefore survive the disaster. The same of course may not be possible for the millions of other species that inhabit the Earth, and that will be on our conscience as a race.

I think we are acting with reckless indifference to our future on planet Earth. At the moment, we have nowhere else to go, but in the long run the human race shouldn't have all its eggs in one basket, or on one planet. I just hope we can avoid dropping the basket before we learn how to escape from Earth. But we are, by nature, explorers. Motivated by curiosity. This is a uniquely human quality. It is this driven curiosity that sent explorers to prove the Earth is not flat and it is the same instinct that sends us to the stars at the speed of thought, urging us to go there in reality. And whenever we make a great new leap, such as the Moon landings, we elevate humanity, bring people and nations together, usher in new discoveries and new technologies. To leave Earth demands a concerted global approach – everyone should join in. We need to rekindle the excitement of the early days of space travel in the 1960s. The technology is almost within our grasp. It is time to explore other solar systems. Spreading out may be the only thing that saves us from ourselves. I am convinced that humans need to leave Earth. If we stay, we risk being annihilated.

●

So, beyond my hope for space exploration, what will the future look like and how might science help us?

The popular picture of science in the future is shown in science-fiction series like *Star Trek*. The producers of *Star Trek* even persuaded me to take part, not that it was difficult.

That appearance was great fun, but I mention it to make a serious point. Nearly all the visions of the future that we have been shown from H. G. Wells onwards have been essentially static. They show a society that is in most cases far in advance of ours, in science, in technology and in political organisation. (The last might not be difficult.) In the period between now and then there must have been great changes, with their accompanying tensions and upsets. But, by the time we are shown the future, science, technology and the organisation of society are supposed to have achieved a level of near-perfection.

I question this picture and ask if we will ever reach a final steady state of science and technology. At no time in the 10,000 years or so since the last Ice Age has the human race been in a state of constant knowledge and fixed technology. There have been a few setbacks, like what we used to call the Dark Ages after the fall of the Roman Empire. But the world's population, which is a measure of our technological ability to preserve life and feed ourselves, has risen steadily, with a few hiccups like the Black Death. In the last 200 years the growth has at times been exponential – and the world population has jumped from 1 billion to about 7.6 billion. Other measures of technological development in recent times are electricity consumption, or the number of scientific articles. They too show near-exponential growth. Indeed, we now have such heightened expectations that some people feel cheated by politicians and scientists because we have not already achieved the Utopian visions of the future. For example, the film *2001: A Space Odyssey* showed us with a base on the Moon and launching a manned, or should I say personned, flight to Jupiter.

There is no sign that scientific and technological development will dramatically slow down and stop in the near future. Certainly not by the time of *Star Trek*, which is only about 350 years away. But the present rate of growth cannot continue for the next millennium. By the year 2600 the world's population would be standing shoulder to shoulder and the electricity consumption would make the Earth glow red hot. If you stacked the new books being published next to each other, at the present rate of production you would have to move at ninety miles an hour just to keep up with the end of the line. Of course, by 2600 new artistic and scientific work will come in electronic forms rather than as physical books and papers. Nevertheless, if the exponential growth continued, there would be ten papers a second in my kind of theoretical physics, and no time to read them.

Clearly the present exponential growth cannot continue indefinitely. So what will happen? One possibility is that we will wipe ourselves out through some disaster such as a nuclear war. Even if we don't destroy ourselves completely there is the possibility that we might descend into a state of brutalism and barbarity, like the opening scene of *Terminator*.

How will we develop in science and technology over the next millennium? This is very difficult to answer. But let me stick my neck out and offer my predictions for the future. I will have some chance of being right about the next hundred years, but the rest of the millennium will be wild speculation.

Our modern understanding of science began about the same time as the European settlement of North America, and by the end of the nineteenth century it seemed that we were about to achieve a complete understanding of the universe in terms of what are now known as classical laws. But, as we have seen, in the twentieth century observations began to show that energy came in discrete packets called quanta and a new kind of theory called quantum mechanics was formulated by Max Planck and others. This presented a completely different picture of reality in which things don't have a single unique history, but have every possible history each with its own probability. When one goes down to the individual particles, the possible particle histories have to include paths that travel faster than light and even paths that go back in time. However, these paths that go back in time are not just like angels dancing on a pin. They have real observational consequences. Even what we think of as empty space is full of particles moving in closed loops in space and time. That is, they move forwards in time on one side of the loop and backwards in time on the other side.

The awkward thing is that because there's an infinite number of points in space and time, there's an infinite number of possible closed loops of particles. And an infinite number of closed loops of particles would have an infinite amount of energy and would curl space and time up to a single point. Even science fiction did not think of anything as odd as this. Dealing with this infinite energy requires some really creative accounting, and much of the work in theoretical physics in the last twenty years has been looking for a theory in which the infinite number of closed loops in space and time cancel each other completely. Only then will we be able to unify quantum theory with Einstein's general relativity and achieve a complete theory of the basic laws of the universe.

What are the prospects that we will discover this complete theory in the next millennium? I would say they were very good, but then I'm an optimist. In 1980 I said I thought there was a 50–50 chance that we would discover a complete unified theory in the next twenty years. We have made some remarkable progress in the period since then, but the final theory seems about the same distance away. Will the Holy

Grail of physics be always just beyond our reach? I think not.

At the beginning of the twentieth century we understood the workings of nature on the scales of classical physics that are good down to about a hundredth of a millimetre. The work on atomic physics in the first thirty years of the century took our understanding down to lengths of a millionth of a millimetre. Since then, research on nuclear and high-energy physics has taken us to length scales that are smaller by a further factor of a billion. It might seem that we could go on forever discovering structures on smaller and smaller length scales. However, there is a limit to this series as with a series of nested Russian dolls. Eventually one gets down to a smallest doll, which can't be taken apart any more. In physics the smallest doll is called the Planck length and is a millimetre divided by a 100,000 billion billion billion. We are not about to build particle accelerators that can probe to distances that small. They would have to be larger than the solar system and they are not likely to be approved in the present financial climate. However, there are consequences of our theories that can be tested by much more modest machines.

It won't be possible to probe down to the Planck length in the laboratory, though we can study the Big Bang to get observational evidence at higher energies and shorter length scales than we can achieve on Earth. However, to a large extent we shall have to rely on mathematical beauty and consistency to find the ultimate theory of everything.

The *Star Trek* vision of the future in which we achieve an advanced but essentially static level may come true in respect of our knowledge of the basic laws that govern the universe. But I don't think we will ever reach a steady state in the uses we make of these laws. The ultimate theory will place no limit on the complexity of systems that we can produce, and it is in this complexity that I think the most important developments of the next millennium will be.

•

By far the most complex systems that we have are our own bodies. Life seems to have originated in the primordial oceans that covered the Earth four billion years ago. How this happened we don't know. It may be that random collisions between atoms built up macromolecules that could reproduce themselves and assemble themselves into more complicated structures. What we do know is that by three and a half billion years ago the highly complicated molecule DNA had emerged. DNA is the basis for all life on Earth. It has a double-helix structure, like a spiral staircase, which was discovered by Francis Crick and James Watson in the Cavendish lab at Cambridge in 1953. The two strands of the double helix are linked by pairs of nitrogenous bases like the treads in a spiral staircase. There are four kinds of nitrogenous bases: cytosine, guanine, adenine and thymine. The order in which the different nitrogenous bases occur along the spiral staircase carries the genetic information that enables the DNA molecule to assemble an organism around it and reproduce itself. As the DNA made copies of itself there would have been occasional errors in the order of the nitrogenous bases along the spiral. In most cases the mistakes in copying would have made the DNA unable to reproduce itself. Such genetic errors, or mutations as they are called, would die out. But in a few cases the error or mutation would increase the chances of the DNA surviving and reproducing. Thus the information content in the sequence of nitrogenous bases would gradually evolve and increase in complexity. This natural selection of mutations was first proposed by another Cambridge man, Charles Darwin, in 1858, though he didn't know the mechanism for it.

Because biological evolution is basically a random walk in the space of all genetic possibilities, it has been very slow. The complexity, or number of bits of information that are coded in DNA, is given roughly by the number of nitrogenous bases in the molecule. Each bit of information can be thought of as the answer to a

What is the biggest threat to the future of this planet?

An asteroid collision would be – a threat against which we have no defence. But the last big such asteroid collision was about sixty-six million years ago and killed the dinosaurs. A more immediate danger is runaway climate change. A rise in ocean temperature would melt the ice caps and cause the release of large amounts of carbon dioxide. Both effects could make our climate like that of Venus, but with a temperature of 250 degrees Celsius.

yes/no question. For the first two billion years or so the rate of increase in complexity must have been of the order of one bit of information every hundred years. The rate of increase of DNA complexity gradually rose to about one bit a year over the last few million years. But now we are at the beginning of a new era in which we will be able to increase the complexity of our DNA without having to wait for the slow process of biological evolution. There has been relatively little change in human DNA in the last 10,000 years. But it is likely that we will be able to redesign it completely in the next thousand. Of course, many people will say that genetic engineering on humans should be banned. But I rather doubt that they will be able to prevent it. Genetic engineering on plants and animals will be allowed for economic reasons, and someone is bound to try it on humans. Unless we have a totalitarian world order, someone will design improved humans somewhere.

Clearly developing improved humans will create great social and political problems with respect to unimproved humans. I'm not advocating human genetic engineering as a good thing, I'm just saying that it is likely to happen in the next millennium, whether we want it or not. This is why I don't believe science fiction like *Star Trek* where people are essentially the same 350 years in the future. I think the human race, and its DNA, will increase its complexity quite rapidly.

In a way, the human race needs to improve its mental and physical qualities if it is to deal with the increasingly complex world around it and meet new challenges like space travel. And it also needs to increase its complexity if biological systems are to keep ahead of electronic ones. At the moment computers have an advantage of speed, but they show no sign of intelligence. This is not surprising because our present computers are less complex than the brain of an earthworm, a species not noted for its intellectual powers. But computers roughly obey a version of Moore's Law, which says that their speed and complexity double every eighteen months. It is one of these exponential growths that clearly cannot continue indefinitely, and indeed it has already begun to slow. However, the rapid pace of improvement will probably continue until computers have a similar complexity to the human brain. Some people say that computers can never show true intelligence, whatever that may be. But it seems to me that if very complicated chemical molecules can operate in humans to make them intelligent, then equally complicated electronic circuits can also make computers act in an intelligent way. And if they are intelligent they can presumably design computers that have even greater complexity and intelligence.

This is why I don't believe the science-fiction picture of an advanced but constant future. Instead, I expect complexity to increase at a rapid rate, in both the biological and the electronic spheres. Not much of this will happen in the next hundred years, which is all we can reliably predict. But by the end of the next millennium, if we get there, the change will be fundamental.

Lincoln Steffens once said, 'I have seen the future and it works.' He was actually talking about the Soviet Union, which we now know didn't work very well. Nevertheless, I think the present world order has a future, but it will be very different.

8

SHOULD WE
COLONISE SPACE?

Why should we go into space? What is the justification for spending all that effort and money on getting a few lumps of moon rock? Aren't there better causes here on Earth? The obvious answer is because it's there, all around us. Not to leave planet Earth would be like castaways on a desert island not trying to escape. We need to explore the solar system to find out where humans could live.

In a way, the situation is like that in Europe before 1492. People might well have argued that it was a waste of money to send Columbus on a wild goose chase. Yet the discovery of the New World made a profound difference to the Old. Just think, we wouldn't have had the Big Mac or KFC. Spreading out into space will have an even greater effect. It will completely change the future of the human race, and maybe determine whether we have any future at all. It won't solve any of our immediate problems on planet Earth, but it will give us a new perspective on them and cause us to look outwards rather than inwards. Hopefully, it will unite us to face the common challenge.

This would be a long-term strategy, and by long term I mean hundreds or even thousands of years. We could have a base on the Moon within thirty years, reach Mars in fifty years and explore the moons of the outer planets in 200 years. By reach, I mean in spacecraft with humans aboard. We have already driven rovers on Mars and landed a probe on Titan, a moon of Saturn, but if we are considering the future of the human race we have to go there ourselves.

Going into space won't be cheap, but it would take only a small proportion of world resources. NASA's budget has remained roughly constant in real terms since the time of the Apollo landings, but it has decreased from 0.3 per cent of US GDP in 1970 to about 0.1 per cent in 2017. Even if we were to increase the international budget twenty times, to make a serious effort to go into space, it would only be a small fraction of world GDP.

There will be those who argue that it would be better to spend our money solving the problems of this planet, like climate change and pollution, rather than wasting it on a possibly fruitless search for a new planet. I'm not denying the importance of fighting climate change and global warming, but we can do that and still spare a quarter of a per cent of world GDP for space. Isn't our future worth a quarter of a per cent?

We thought space was worth a big effort in the 1960s. In 1962, President Kennedy committed the US to landing a man on the Moon by the end of the decade. On 20 July 1969, Buzz Aldrin and Neil Armstrong landed on the surface of the Moon. It changed the future of the human race. I was twenty-seven at the time, a researcher at Cambridge, and I missed it. I was at a meeting on singularities in Liverpool and listening to a lecture by René Thom on catastrophe theory when the landing took place. There was no catch-up TV in those days, and we didn't have a television, but my son aged two described it to me.

The space race helped to create a fascination with science and accelerated our technological progress. Many of today's scientists were inspired to go into science as a result of the Moon landings, with the aim of understanding more about ourselves and our place in the universe. It gave us new perspectives on our world, prompting us to consider the planet as a whole. However, after the last Moon landing in 1972, with no future plans for further manned space flight, public interest in space declined. This went along with a general disenchantment with science in the West, because although it had brought great benefits it had not solved the social problems that increasingly occupied public attention.

A new crewed space flight programme would do a lot to restore public enthusiasm for space and for science generally. Robotic missions are much cheaper and may provide more scientific information, but they don't catch the public imagination in the same way. And they don't spread the human race into space, which I'm arguing should be our long-term strategy. A goal of a base on the Moon by 2050, and of a manned landing on Mars by 2070, would reignite the space programme, and give it a sense of purpose, in the same way that President Kennedy's Moon target did in the 1960s. In late 2017, Elon Musk announced SpaceX plans for a lunar base and a Mars mission by 2022, and President Trump signed a space policy directive refocusing NASA on exploration and discovery, so perhaps we'll get there even sooner.

A new interest in space would also increase the public standing of science generally. The low esteem in which science and scientists are held is having serious consequences. We live in a society that is increasingly governed by science and technology, yet fewer and fewer young people want to go into science. A new and ambitious space programme would excite the young and stimulate them into entering a wide range of sciences, not just astrophysics and space science.

The same is true for me. I have always dreamed of space flight. But for so many years I thought it was just that, a dream. Confined to Earth and in a wheelchair, how could I experience the majesty of space except through imagination and my work in theoretical physics. I never thought I would have the opportunity to see our beautiful planet from space or gaze out into the infinity beyond. This was the domain of astronauts, the lucky few who get to experience the wonder and thrill of space flight. But I had not factored in the energy and enthusiasm of individuals whose mission it is to take that first step in venturing outside Earth. And in 2007 I was fortunate enough to go on a zero-gravity flight and experience weightlessness for the first time. It only lasted for four minutes, but it was amazing. I could have gone on and on.

I was quoted at the time as saying that I feared the human race is not going to have a future if we don't go into space. I believed it then, and I believe it still. And I hope I demonstrated then that anyone can take part in space travel. I believe it is up to scientists like me, together with innovative commercial entrepreneurs, to do all we can to promote the excitement and wonder of space travel.

But can humans exist for long periods away from the Earth? Our experience with the ISS, the International Space Station, shows that it is possible for human beings to survive for many months away from planet Earth. However, the zero gravity of orbit causes a number of undesirable physiological changes, including a weakening of the bones, as well as creating practical problems with liquids and so on. One would therefore want any long-term base for human beings to be on a planet or moon. By digging into the surface, one would get thermal insulation, and protection from meteors and cosmic rays. The planet or moon could also serve as a source of the raw materials that would be needed if the extra-terrestrial community was to be self-sustaining, independent of Earth.

What are the possible sites of a human colony in the solar system? The most obvious is the Moon. It is close by and relatively easy to reach. We have already landed on it, and driven across it in a buggy. On the other hand, the Moon is small, and without atmosphere, or a magnetic field to deflect the solar-radiation particles, like on Earth. There is no liquid water, although there may be ice in the craters at the North and South Poles. A colony on the Moon could use this as a source of oxygen, with power provided by nuclear energy or solar panels. The Moon could be a base for travel to the rest of the solar system.

Mars is the obvious next target. It is half as far again as the Earth from the Sun, and so receives half the warmth. It once had a magnetic field, but it decayed four billion years ago, leaving Mars without protection from solar radiation. This stripped Mars of most of its atmosphere, leaving it with only 1 per cent of the pressure of the Earth's atmosphere. However, the pressure must have been higher in the past, because we see what appear to be run-off channels and dried-up lakes. Liquid water cannot exist on the surface of Mars now. It would vaporise in the near-vacuum. This suggests that Mars had a warm wet period, during which life might have appeared, either spontaneously or through panspermia (that is, brought from somewhere else in the universe). There is no sign of life on Mars now, but if we found evidence that life had once existed it would indicate that the probability of life developing on a suitable planet was fairly high. We must be careful, though, that we don't confuse the issue by contaminating the planet with life from Earth. Similarly, we must be very careful not to bring back any Martian life. We would have no resistance to it, and it might wipe out life on Earth.

NASA has sent a large number of spacecraft to Mars, starting with Mariner 4 in 1964. It has surveyed the planet with a number of orbiters, the latest being the Mars reconnaissance orbiter. These orbiters have revealed deep gulleys and the highest mountains in the solar system. NASA has also landed a number of probes on the surface of Mars, most recently the two Mars rovers. These have sent back pictures of a dry desert landscape. Like on the Moon, water and oxygen might be obtainable from polar ice. There has been volcanic activity on Mars. This would have brought minerals and metals to the surface, which a colony could use.

The Moon and Mars are the most suitable sites for space colonies in the solar system. Mercury and Venus are too hot, while Jupiter and Saturn are gas giants with no solid surface. The moons of Mars are very small and have no advantages over Mars itself. Some of the moons of Jupiter and Saturn might be possible. Europa, a moon of Jupiter, has a frozen ice surface. But there may be liquid water under the surface in which life could have developed. How can we find out? Do we have to land on Europa and drill a hole?

Titan, a moon of Saturn, is larger and more massive than our Moon and has a dense atmosphere. The Cassini–Huygens mission of NASA and the European Space Agency has landed a probe on Titan which has sent back pictures of the surface. However, it is very cold, being so far from the Sun, and I wouldn't fancy living next to a lake of liquid methane.

But what about boldly going beyond the solar system? Our observations indicate that a significant fraction of stars have planets around them. So far, we can detect only giant planets, like Jupiter and Saturn, but it is reasonable to assume that they will be accompanied by smaller, Earth-like planets. Some of these will lie in the Goldilocks zone, where the distance from the star is in the right range for liquid water to exist on their surface. There are around a thousand stars within thirty light years of Earth. If 1 per cent of these have Earth-sized planets in the Goldilocks zone, we have ten candidate New Worlds.

Take Proxima b, for example. This exoplanet, which is the closest to Earth but still four and a half light years away, orbits the star Proxima Centauri within the solar system Alpha Centauri, and recent research indicates that it has some similarities to Earth.

Travelling to these candidate worlds isn't possible perhaps with today's technology, but by using our imagination we can make interstellar travel a long-term aim – in the next 200 to 500 years. The speed at which we can send a rocket is governed by two things, the speed of the exhaust and the fraction of its mass that the rocket loses as it accelerates. The exhaust speed of chemical rockets, like the ones we have used so far, is about three kilometres per second. This accelerates rockets ever faster and by jettisoning some of their mass they increase their speed even more. According to NASA, it would take as little as 260 days to reach Mars, give or take ten days, with some NASA scientists predicting as little as 130 days. But it would take three million years to get to the nearest star system. To go

BRIEF ANSWERS TO THE BIG QUESTIONS

faster would require a much higher exhaust speed than chemical rockets can provide, that of light itself. A powerful beam of light from the rear could drive the spaceship forward. Nuclear fusion could provide 1 per cent of the spaceship's mass energy, which would accelerate it to a tenth of the speed of light. Beyond that, we would need either matter–antimatter annihilation or some completely new form of energy. In fact, the distance to Alpha Centauri is so great that to reach it in a human lifetime a spacecraft would have to carry fuel with roughly the mass of all the stars in the galaxy. In other words, with current technology interstellar travel is utterly impractical. Alpha Centauri can never become a holiday destination.

We have a chance to change that, thanks to imagination and ingenuity. In 2016 I joined with the entrepreneur Yuri Milner to launch Breakthrough Starshot, a long-term research and development programme aimed at making interstellar travel a reality. If we succeed, we will send a probe to Alpha Centauri within the lifetime of people alive today. But I will return to this shortly.

How do we start this journey? So far, our explorations have been limited to our local cosmic neighbourhood. Forty years on, our most intrepid explorer, Voyager, has just made it to interstellar space. Its speed, eleven miles a second, means it would take about 70,000 years to reach Alpha Centauri. This constellation is 4.37 light years away, twenty-five trillion miles. If there are beings alive on Alpha Centauri today, they remain blissfully ignorant of the rise of Donald Trump.

It is clear we are entering a new space age. The first private astronauts will be pioneers, and the first flights will be hugely expensive, but over time it is my hope that space flight will become within the reach of far more of the Earth's population. Taking more and more passengers into space will bring new meaning to our place on Earth and to our responsibilities as its stewards, and it will help us to recognise our place and future in the cosmos – which is where I believe our ultimate destiny lies.

Breakthrough Starshot is a real opportunity for man to make early forays into outer space, with a view to probing and weighing the possibilities of colonisation. It is a proof-of-concept mission and works on three concepts: miniaturised spacecraft, light propulsion and phase-locked lasers. The Star Chip, a fully functional space probe reduced to a few centimetres in size, will be attached to a light sail. Made from metamaterials, the light sail weighs no more than a few grams. It is envisaged that a thousand Star Chips and light sails, the nanocraft, will be sent into orbit. On the ground, an array of lasers at the kilometre scale will combine into a single, very powerful light beam. The beam is fired through the atmosphere, striking the sails in space with tens of gigawatts of power.

The idea behind this innovation is that the nanocraft ride on the light beam much as Einstein dreamed about riding a light beam at the age of sixteen. Not quite to the speed of light, but to a fifth of it, or 100 million miles an hour. Such a system could reach Mars in less than an hour, reach Pluto in days, pass Voyager in under a week and reach Alpha Centauri in just over twenty years. Once there, the nanocraft could image any planets discovered in the system, test for magnetic fields and organic molecules and send the data back to Earth in another laser beam. This tiny signal would be received by the same array of dishes that were used to transit the launch beam, and return is estimated to take about four years. Importantly, the Star Chips' trajectories may include a fly-by of Proxima b, the Earth-sized planet that is in the habitable zone of its host star, in Alpha Centauri. In 2017, Breakthrough e European Southern Observatory joined forces her a search for habitable planets in Alpha i.

There are secondary targets for Breakthrough Starshot. It would explore the solar system and detect asteroids that cross the path of Earth's orbit around the Sun. In addition, the German physicist Claudius Gros

The era of civilian space travel is coming. What do you think it means to us?

I look forward to space travel. I would be one of the first to buy a ticket. I expect that within the next hundred years we will be able to travel anywhere in the solar system, except maybe the outer planets. But travel to the stars will take a bit longer. I reckon in 500 years, we will have visited some of the nearby stars. It won't be like *Star Trek*. We won't be able to travel at warp speed. So a round trip will take at least ten years and probably much longer.

has proposed that this technology may also be used to establish a biosphere of unicellular microbes on otherwise only transiently habitable exoplanets.

So far, so possible. However, there are major challenges. A laser with a gigawatt of power would provide only a few newtons of thrust. But the nanocraft compensate for this by having a mass of only a few grams. The engineering challenges are immense. The nanocraft must survive extreme acceleration, cold, vacuum and protons, as well as collisions with junk such as space dust. In addition, focusing a set of lasers totalling 100 gigawatts on the solar sails will be difficult due to atmospheric turbulence. How do we combine hundreds of lasers through the motion of the atmosphere, how do we propel the nanocraft without incinerating them and how do we aim them in the right direction? Then we would need to keep the nanocraft functioning for twenty years in the frozen void, so they can send back signals across four light years. But these are engineering problems, and engineers' challenges tend, eventually, to be solved. As it progresses into a mature technology, other exciting missions can be envisaged. Even with less powerful laser arrays, journey times to other planets, to the outer solar system or to interstellar space could be vastly reduced.

Of course, this would not be human interstellar

travel, even if it could be scaled up to a crewed vessel. It would be unable to stop. But it would be the moment when human culture goes interstellar, when we finally reach out into the galaxy. And if Breakthrough Starshot should send back images of a habitable planet orbiting our closest neighbour, it could be of immense importance to the future of humanity.

In conclusion, I return to Einstein. If we find a planet in the Alpha Centauri system, its image, captured by a camera travelling at a fifth of light speed, will be slightly distorted due to the effects of special relativity. It would be the first time a spacecraft has flown fast enough to see such effects. In fact, Einstein's theory is central to the whole mission. Without it we would have neither lasers nor the ability to perform the calculations necessary for guidance, imaging and data transmission over twenty-five trillion miles at a fifth of light speed.

We can see a pathway between that sixteen-year-old boy dreaming of riding on a light beam and our own dream, which we are planning to turn into a reality, of riding our own light beam to the stars. We are standing at the threshold of a new era. Human colonisation on other planets is no longer science fiction. It can be science fact. The human race has existed as a separate species for about two million years. Civilisation began about 10,000 years ago, and the rate of development has been steadily increasing. If humanity is to continue for another million years, our future lies in boldly going where no one else has gone before.

I hope for the best. I have to. We have no other option.

9

WILL ARTIFICIAL INTELLIGENCE OUTSMART US?

Intelligence is central to what it means to be human. Everything that civilisation has to offer is a product of human intelligence.

DNA passes the blueprints of life between generations. Ever more complex life forms input information from sensors such as eyes and ears and process the information in brains or other systems to figure out how to act and then act on the world, by outputting information to muscles, for example. At some point during our 13.8 billion years of cosmic history, something beautiful happened. This information processing got so intelligent that life forms became conscious. Our universe has now awoken, becoming aware of itself. I regard it a triumph that we, who are ourselves mere stardust, have come to such a detailed understanding of the universe in which we live.

I think there is no significant difference between how the brain of an earthworm works and how a computer computes. I also believe that evolution implies there can be no qualitative difference between the brain of an earthworm and that of a human. It therefore follows that computers can, in principle, emulate human intelligence, or even better it. It's clearly possible for something to acquire higher intelligence than its ancestors: we evolved to be smarter than our ape-like ancestors, and Einstein was smarter than his parents.

If computers continue to obey Moore's Law, doubling their speed and memory capacity every eighteen months, the result is that computers are likely to overtake humans in intelligence at some point in the next hundred years. When an artificial intelligence (AI) becomes better than humans at AI design, so that it can recursively improve itself without human help, we may face an intelligence explosion that ultimately results in machines whose intelligence exceeds ours by more than ours exceeds that of snails. When that happens, we will need to ensure that the computers have goals aligned with ours. It's tempting to dismiss the notion of highly intelligent machines as mere science fiction, but this would be a mistake, and potentially our worst mistake ever.

For the last twenty years or so, AI has been focused on the problems surrounding the construction of intelligent agents, systems that perceive and act in a particular environment. In this context, intelligence is related to statistical and economic notions of rationality – that is, colloquially, the ability to make good decisions, plans or inferences. As a result of this recent work, there has been a large degree of integration and cross-fertilisation among AI, machine-learning, statistics, control theory, neuroscience and other fields. The establishment of shared theoretical frameworks, combined with the availability of data and processing power, has yielded remarkable successes in various component tasks, such as speech recognition, image classification, autonomous vehicles, machine translation, legged locomotion and question-answering systems.

As development in these areas and others moves from laboratory research to economically valuable technologies, a virtuous cycle evolves, whereby even small improvements in performance are worth large sums of money, prompting further and greater investments in research. There is now a broad consensus that AI research is progressing steadily and that its impact on society is likely to increase. The potential benefits are huge; we cannot predict what we might achieve when this intelligence is magnified by the tools AI may provide. The eradication of disease and poverty is possible. Because of the great potential of AI, it is important to research how to reap its benefits while avoiding potential pitfalls. Success in creating AI would be the biggest event in human history.

Unfortunately, it might also be the last, unless we learn how to avoid the risks. Used as a toolkit, AI can augment our existing intelligence to open up advances in every area of science and society. However, it will also bring dangers. While primitive forms of artificial intelligence developed so far have proved very useful, I fear the consequences of creating something that can match or surpass humans. The concern is that AI would take off on its own and redesign itself at an ever-increasing rate. Humans, who are limited by slow biological evolution, couldn't compete and would be superseded. And in the future AI could develop a will of its own, a will that is in conflict with ours. Others believe that humans can command the rate of technology for a decently long time, and that the potential of AI to solve many of the world's problems will be realised. Although I am well known as an optimist regarding the human race, I am not so sure.

In the near term, for example, world militaries are considering starting an arms race in autonomous-weapon systems that can choose and eliminate their own targets. While the UN is debating a treaty banning such weapons, autonomous-weapons proponents usually forget to ask the most important question. What is the likely end-point of an arms race and is that desirable for the human race? Do we really want cheap AI weapons to become the Kalashnikovs of tomorrow, sold to criminals and terrorists on the black market? Given concerns about our ability to maintain long-term control of ever more advanced AI systems, should we arm them and turn over our defence to them? In 2010, computerised trading systems created the stock-market Flash Crash; what would a computer-triggered crash look like in the defence arena? The best time to stop the autonomous-weapons arms race is now.

In the medium term, AI may automate our jobs, to bring both great prosperity and equality. Looking further ahead, there are no fundamental limits to what can be achieved. There is no physical law precluding particles from being organised in ways that perform even more advanced computations than the arrangements of particles in human brains. An explosive transition is possible, although it may play out differently than in the movies. As mathematician Irving Good realised in 1965, machines with superhuman intelligence could repeatedly improve their design even further, in what science-fiction writer Vernor Vinge called a technological singularity. One can imagine such technology outsmarting financial markets, out-inventing human researchers, out-manipulating human leaders and potentially subduing us with weapons we cannot even understand. Whereas the short-term impact of AI depends on who controls it, the long-term impact depends on whether it can be controlled at all.

In short, the advent of super-intelligent AI would be either the best or the worst thing ever to happen to humanity. The real risk with AI isn't malice but competence. A super-intelligent AI will be extremely good at accomplishing its goals, and if those goals aren't aligned with ours we're in trouble. You're probably not an evil ant-hater who steps on ants out of malice, but if you're in charge of a hydroelectric green-energy project and there's an anthill in the region to be flooded, too bad for the ants. Let's not place humanity in the position of those ants. We should plan ahead. If a superior alien civilisation sent us a text message saying, 'We'll arrive in a few decades', would we just reply, 'OK, call us when you get here, we'll leave the lights on'? Probably not, but this is more or less what has happened with AI. Little serious research has been devoted to these issues outside a few small non-profit institutes.

Fortunately, this is now changing. Technology pioneers Bill Gates, Steve Wozniak and Elon Musk have echoed my concerns, and a healthy culture of risk assessment and awareness of societal implications is beginning to take root in the AI community. In January 2015, I, along with Elon Musk and many AI experts, signed an open letter on artificial intelligence, calling for serious research into its impact on society. In the past, Elon Musk has warned that superhuman artificial intelligence is capable of providing incalculable benefits, but if deployed incautiously will have an adverse effect on the human race. He and I sit on the scientific advisory board for the Future of Life Institute, an organisation working to mitigate existential risks facing humanity, and which drafted the open letter. This called for concrete research on how we could prevent potential problems while also reaping the potential benefits AI offers us, and is designed to get AI researchers and developers to pay more attention to AI safety. In addition, for policymakers and the general public the letter was meant to be informative but not alarmist. We think it is very important that everybody knows that AI researchers are seriously thinking about these concerns and ethical issues. For example, AI has the potential to eradicate disease and poverty, but researchers must work to create AI that can be controlled.

In October 2016, I also opened a new centre in Cambridge, which will attempt to tackle some of the open-ended questions raised by the rapid pace of development in AI research. The Leverhulme Centre for the Future of Intelligence is a multi-disciplinary institute, dedicated to researching the future of intelligence as crucial to the future of our civilisation and our species. We spend a great deal of time studying history, which, let's face it, is mostly the history of stupidity. So it's a welcome change that people are studying instead the future of intelligence. We are aware of the potential dangers, but perhaps with the tools of this new technological revolution we will even be able to undo some of the damage done to the natural world by industrialisation.

Recent developments in the advancement of AI include a call by the European Parliament for drafting a set of regulations to govern the creation of robots and AI. Somewhat surprisingly, this includes a form of electronic personhood, to ensure the rights and responsibilities for the most capable and advanced AI. A European Parliament spokesman has commented that, as a growing number of areas in our daily lives are increasingly affected by robots, we need to ensure that robots are, and will remain, in the service of humans. A report presented to the Parliament declares that the world is on the cusp of a new industrial robot revolution. It examines whether or not providing legal rights for robots as electronic persons, on a par with the legal definition of corporate personhood, would be permissible. But it stresses that at all times researchers and designers should ensure all robotic design incorporates a kill switch.

This didn't help the scientists on board the space-ship with Hal, the malfunctioning robotic computer in

Stanley Kubrick's *2001: A Space Odyssey*, but that was fiction. We deal with fact. Lorna Brazell, a consultant at the multinational law firm Osborne Clarke, says in the report that we don't give whales and gorillas personhood, so there is no need to jump at robotic personhood. But the wariness is there. The report acknowledges the possibility that within a few decades AI could surpass human intellectual capacity and challenge the human–robot relationship.

By 2025, there will be about thirty mega-cities, each with more than ten million inhabitants. With all those people clamouring for goods and services to be delivered whenever they want them, can technology help us keep pace with our craving for instant commerce? Robots will definitely speed up the online retail process. But to revolutionise shopping they need to be fast enough to allow same-day delivery on every order.

Opportunities for interacting with the world, without having to be physically present, are increasing rapidly. As you can imagine, I find that appealing, not least because city life for all of us is so busy. How many times have you wished you had a double who could share your workload? Creating realistic digital surrogates of ourselves is an ambitious dream, but the latest technology suggests that it may not be as far-fetched an idea as it sounds.

When I was younger, the rise of technology pointed

to a future where we would all enjoy more leisure time. But in fact the more we can do, the busier we become. Our cities are already full of machines that extend our capabilities, but what if we could be in two places at once? We're used to automated voices on phone systems and public announcements. Now inventor Daniel Kraft is investigating how we can replicate ourselves visually. The question is, how convincing can an avatar be?

Interactive tutors could prove useful for massive open online courses (MOOCs) and for entertainment. It could be really exciting – digital actors that would be forever young and able to perform otherwise impossible feats. Our future idols might not even be real.

How we connect with the digital world is key to the progress we'll make in the future. In the smartest cities, the smartest homes will be equipped with devices that are so intuitive they'll be almost effortless to interact with.

When the typewriter was invented, it liberated the way we interact with machines. Nearly 150 years later and touch screens have unlocked new ways to communicate with the digital world. Recent AI landmarks, such as self-driving cars, or a computer winning at the game of Go, are signs of what is to come. Enormous levels of investment are pouring into this technology, which already forms a major part of our lives. In the coming decades it will permeate every aspect of our

Why are we so worried about artificial intelligence? Surely humans are always able to pull the plug?

People asked a computer, 'Is there a God?'
And the computer said, 'There is now,'
and fused the plug.

society, intelligently supporting and advising us in many areas including healthcare, work, education and science. The achievements we have seen so far will surely pale against what the coming decades will bring, and we cannot predict what we might achieve when our own minds are amplified by AI.

Perhaps with the tools of this new technological revolution we can make human life better. For instance, researchers are developing AI that would help reverse paralysis in people with spinal-cord injuries. Using silicon chip implants and wireless electronic interfaces between the brain and the body, the technology would allow people to control their body movements with their thoughts.

I believe the future of communication is brain–computer interfaces. There are two ways: electrodes on the skull and implants. The first is like looking through frosted glass, the second is better but risks infection. If we can connect a human brain to the internet it will have all of Wikipedia as its resource.

The world has been changing even faster as people, devices and information are increasingly connected to each other. Computational power is growing and quantum computing is quickly being realised. This will revolutionise artificial intelligence with exponentially faster speeds. It will advance encryption. Quantum computers will change everything, even human biology.

There is already one technique to edit DNA precisely, called CRISPR. The basis of this genome-editing technology is a bacterial defence system. It can accurately target and edit stretches of genetic code. The best intention of genetic manipulation is that modifying genes would allow scientists to treat genetic causes of disease by correcting gene mutations. There are, however, less noble possibilities for manipulating DNA. How far we can go with genetic engineering will become an increasingly urgent question. We can't see the possibilities of curing motor neurone diseases – like my ALS – without also glimpsing its dangers.

Intelligence is characterised as the ability to adapt to change. Human intelligence is the result of generations of natural selection of those with the ability to adapt to changed circumstances. We must not fear change. We need to make it work to our advantage.

We all have a role to play in making sure that we, and the next generation, have not just the opportunity but the determination to engage fully with the study of science at an early level, so that we can go on to fulfil our potential and create a better world for the whole human race. We need to take learning beyond a theoretical discussion of how AI should be and to make sure we plan for how it can be. We all have the potential to push the boundaries of what is accepted, or expected, and to think big. We stand on the threshold of a brave new world. It is an exciting, if precarious, place to be, and we are the pioneers.

When we invented fire, we messed up repeatedly, then invented the fire extinguisher. With more powerful technologies such as nuclear weapons, synthetic biology and strong artificial intelligence, we should instead plan ahead and aim to get things right the first time, because it may be the only chance we will get. Our future is a race between the growing power of our technology and the wisdom with which we use it. Let's make sure that wisdom wins.

10

HOW DO WE
SHAPE THE FUTURE?

A century ago, Albert Einstein revolutionised our understanding of space, time, energy and matter. We are still finding awesome confirmations of his predictions, like the gravitational waves observed in 2016 by the LIGO experiment. When I think about ingenuity, Einstein springs to mind. Where did his ingenious ideas come from? A blend of qualities, perhaps: intuition, originality, brilliance. Einstein had the ability to look beyond the surface to reveal the underlying structure. He was undaunted by common sense, the idea that things must be the way they seemed. He had the courage to pursue ideas that seemed absurd to others. And this set him free to be ingenious, a genius of his time and every other.

A key element for Einstein was imagination. Many of his discoveries came from his ability to reimagine the universe through thought experiments. At the age of sixteen, when he visualised riding on a beam of light, he realised that from this vantage light would appear as a frozen wave. That image ultimately led to the theory of special relativity.

One hundred years later, physicists know far more about the universe than Einstein did. Now we have greater tools for discovery, such as particle accelerators, supercomputers, space telescopes and experiments such as the LIGO lab's work on gravitational waves. Yet imagination remains our most powerful attribute. With it, we can roam anywhere in space and time. We can witness nature's most exotic phenomena while driving in a car, snoozing in bed or pretending to listen to someone boring at a party.

As a boy, I was passionately interested in how things worked. In those days, it was more straightforward to take something apart and figure out the mechanics. I was not always successful in reassembling toys I had pulled to pieces, but I think I learned more than a boy or girl today would, if he or she tried the same trick on a smartphone.

My job now is still to figure out how things work, only the scale has changed. I don't destroy toy trains any more. Instead, I try to figure out how the universe works, using the laws of physics. If you know how something works, you can control it. It sounds so simple when I say it like that! It is an absorbing and complex endeavour that has fascinated and thrilled me throughout my adult life. I have worked with some of the greatest scientists in the world. I have been lucky to be alive through what has been a glorious time in my chosen field, cosmology, the study of the origins of the universe.

The human mind is an incredible thing. It can conceive of the magnificence of the heavens and the intricacies of the basic components of matter. Yet for each mind to achieve its full potential, it needs a spark. The spark of enquiry and wonder.

Often that spark comes from a teacher. Allow me to explain. I wasn't the easiest person to teach, I was slow to learn to read and my handwriting was untidy. But when I was fourteen my teacher at my school in St Albans, Dikran Tahta, showed me how to harness my energy and encouraged me to think creatively about mathematics. He opened my eyes to maths as the blueprint of the universe itself. If you look behind every exceptional person there is an exceptional teacher. When each of us thinks about what we can do in life, chances are we can do it because of a teacher.

However, education and science and technology research are endangered now more than ever before. Due to the recent global financial crisis and austerity measures, funding is being significantly cut to all areas of science, but in particular the fundamental sciences have been badly affected. We are also in danger of becoming culturally isolated and insular, and increasingly remote from where progress is being made. At the level of research, the exchange of people across borders enables skills to transfer more quickly and brings new people with different ideas, derived from their different backgrounds. This can easily make for progress where now this progress will be harder. Unfortunately, we cannot go back in time. With Brexit and Trump now exerting new forces in relation to immigration and the development of education, we are witnessing a global revolt against experts, which includes scientists. So what can we do to secure the future of science and technology education?

I return to my teacher, Mr Tahta. The basis for the future of education must lie in schools and inspiring teachers. But schools can only offer an elementary framework where sometimes rote-learning, equations and examinations can alienate children from science. Most people respond to a qualitative, rather than a quantitative, understanding, without the need for complicated equations. Popular science books and articles can also put across ideas about the way we live. However, only a small percentage of the population read even the most successful books. Science documentaries and films reach a mass audience, but it is only one-way communication.

When I started out in the field in the 1960s, cosmology was an obscure and cranky branch of scientific study. Today, through theoretical work and experimental triumphs such as the Large Hadron Collider and the discovery of the Higgs boson, cosmology has opened the universe up to us. There are big questions still to answer and much work lies ahead. But we know more now and have achieved more in this relatively short space of time than anyone could have imagined.

But what lies ahead for those who are young now? I can say with confidence that their future will depend more on science and technology than any previous generation's has done. They need to know about science more than any before them because it is part of their daily lives in an unprecedented way.

Without speculating too wildly, there are trends we can see and emerging problems that we know must be dealt with, now and into the future. Among the problems I count global warming, finding space and resources for the massive increase in the Earth's human population, rapid extinction of other species, the need to develop renewable energy sources, the degradation of the oceans, deforestation and epidemic diseases – just to name a few.

There are also the great inventions of the future, which will revolutionise the ways we live, work, eat, communicate and travel. There is such enormous scope for innovation in every area of life. This is exciting. We could be mining rare metals on the Moon, establishing a human outpost on Mars and finding cures and treatments for conditions which currently offer no hope. The huge questions of existence still remain unanswered – how did life begin on Earth? What is consciousness? Is there anyone out there or are we alone in the universe? These are questions for the next generation to work on.

Some think that humanity today is the pinnacle of evolution, and that this is as good as it gets. I disagree. There ought to be something very special about the boundary conditions of our universe, and what can be more special than that there is no boundary. And there should be no boundary to human endeavour. We have two options for the future of humanity as I see it: first, the exploration of space for alternative planets on which to live, and second, the positive use of artificial intelligence to improve our world.

The Earth is becoming too small for us. Our physical resources are being drained at an alarming rate. Mankind has presented our planet with the disastrous gifts of climate change, pollution, rising temperatures, reduction of the polar ice caps, deforestation and decimation of animal species. Our population, too, is increasing at an alarming rate. Faced with these figures, it is clear this near-exponential population growth cannot continue into the next millennium.

Another reason to consider colonising another planet is the possibility of nuclear war. There is a theory that says the reason we have not been contacted by extra-terrestrials is that when a civilisation reaches our stage of development it becomes unstable and destroys itself. We now have the technological power to destroy every living creature on Earth. As we have seen in recent events in North Korea, this is a sobering and worrying thought.

But I believe we can avoid this potential for Armageddon, and one of the best ways for us to do this is to move out into space and explore the potential for humans to live on other planets.

The second development which will impact on the future of humanity is the rise of artificial intelligence.

Artificial intelligence research is now progressing rapidly. Recent landmarks such as self-driving cars, a computer winning the game of Go and the arrival of digital personal assistants Siri, Google Now and Cortana are merely symptoms of an IT arms race, fuelled by unprecedented investments and building on an increasingly mature, theoretical foundation. Such achievements will probably pale against what the coming decades will bring.

But the advent of super-intelligent AI would be either

the best or the worst thing ever to happen to humanity. We cannot know if we will be infinitely helped by AI, or ignored by it and sidelined, or conceivably destroyed by it. As an optimist, I believe that we can create AI for the good of the world, that it can work in harmony with us. We simply need to be aware of the dangers, identify them, employ the best possible practice and management and prepare for its consequences well in advance.

Technology has had a huge impact on my life. I speak through a computer. I have benefited from assisted technology to give me a voice that my illness has taken away. I was lucky to lose my voice at the beginning of the personal computing age. Intel has been supporting me for over twenty-five years, allowing me to do what I love every day. Over these years the world, and technology's impact on it, has changed dramatically. Technology has changed the way we all live our lives, from communication to genetic research, to access to information, and much, much more. As technology has got smarter, it has opened doors to possibilities that I didn't ever predict. The technology that is now being developed to support the disabled is leading the way in breaking down the communication barriers which once stood in the way. It is often a proving ground for the technology of the future. Voice to text, text to voice, home automation, drive by wire, even the Segway, were developed for the disabled, years before they were in everyday use. These technological achievements are due to the spark of fire within ourselves, the creative force. This creativity can take many forms, from physical achievement to theoretical physics.

But so much more will happen. Brain interfaces could make this means of communication – used by more and more people – quicker and more expressive. I now use Facebook – it allows me to speak directly to my friends and followers worldwide so they can keep up with my latest theories and see pictures from my travels. It also means I can see what my children are really up to, rather than what they tell me they are doing.

In the same way that the internet, our mobile phones, medical imaging, satellite navigation and social networks would have been incomprehensible to the society of only a few generations ago, our future world will be equally transformed in ways we are only beginning to conceive. Information on its own will not take us there, but its intelligent and creative use will.

There is so much more to come and I hope that this prospect offers great inspiration to schoolchildren today. But we have a role to play in making sure this generation of children have not just the opportunity but the wish to engage fully with the study of science at an early level so that they can go on to fulfil their potential and create a better world for the whole human race. And I believe the future of learning and education is the internet. People can answer back and interact. In a way, the internet connects us all together like the neurons in a giant brain. And with such an IQ, what cannot we be capable of?

When I was growing up it was still acceptable – not to me but in social terms – to say that one was not interested in science and did not see the point in bothering with it. This is no longer the case. Let me be clear. I am not promoting the idea that all young people should grow up to be scientists. I do not see that as an ideal situation, as the world needs people with a wide variety of skills. But I am advocating that all young people should be familiar with and confident around scientific subjects, whatever they choose to do. They need to be scientifically literate, and inspired to engage with developments in science and technology in order to learn more.

A world where only a tiny super-elite are capable of understanding advanced science and technology and its applications would be, to my mind, a dangerous and limited one. I seriously doubt whether long-range beneficial projects such as cleaning up the oceans or curing diseases in the developing world would be given priority. Worse, we could find that technology is used against us and that we might have no power to stop it.

I don't believe in boundaries, either for what we can

What world-changing idea, small or big, would you like to see implemented by humanity?

This is easy. I would like to see the development of fusion power to give an unlimited supply of clean energy, and a switch to electric cars. Nuclear fusion would become a practical power source and would provide us with an inexhaustible supply of energy, without pollution or global warming.

do in our personal lives or for what life and intelligence can accomplish in our universe. We stand at a threshold of important discoveries in all areas of science. Without doubt, our world will change enormously in the next fifty years. We will find out what happened at the Big Bang. We will come to understand how life began on Earth. We may even discover whether life exists elsewhere in the universe. While the chances of communicating with an intelligent extra-terrestrial species may be slim, the importance of such a discovery means we must not give up trying. We will continue to explore our cosmic habitat, sending robots and humans into space. We cannot continue to look inwards at ourselves on a small and increasingly polluted and overcrowded planet. Through scientific endeavour and technological innovation, we must look outwards to the wider universe, while also striving to fix the problems on Earth. And I am optimistic that we will ultimately create viable habitats for the human race on other planets. We will transcend the Earth and learn to exist in space.

This is not the end of the story, but just the beginning of what I hope will be billions of years of life flourishing in the cosmos.

And one final point – we never really know where the next great scientific discovery will come from, nor who will make it. Opening up the thrill and wonder of scientific discovery, creating innovative and accessible ways to reach out to the widest young audience possible, greatly increases the chances of finding and inspiring the new Einstein. Wherever she might be.

So remember to look up at the stars and not down at your feet. Try to make sense of what you see and wonder about what makes the universe exist. Be curious. And however difficult life may seem, there is always something you can do and succeed at. It matters that you don't just give up. Unleash your imagination. Shape the future.

Afterword

Lucy Hawking

On the bleak greyness of a Cambridge spring day, we set off in a cortège of black cars towards Great St Mary's Church, the university church where distinguished academics by tradition have their funeral services. Out of term, the streets seemed muted. Cambridge looked empty, not even a wandering tourist in sight. The only spikes of colour came from the blue flashing lights of the police motorcycle outriders, guarding the hearse with my father's coffin in it, stopping the sparse traffic as we went.

And then we turned left. And saw the crowds, massed along one of the most recognisable streets in the world, King's Parade, the heart of Cambridge itself. I have never seen so many people so silent. With banners, flags, cameras and mobile phones held aloft, the huge numbers of people lining the streets stood in quiet respect as the head porter of Gonville and Caius, my father's Cambridge college, dressed ceremonially in his bowler hat and carrying an ebony cane, walked solemnly along the street to meet the hearse and walk it to the church.

My aunt squeezed my hand as we both burst into tears. 'He would have loved this,' she whispered to me.

Since my father died, there has been so much he would have loved, so much I wish he could have known. I wish he could have seen the extraordinary outpouring of affection towards him, coming from all around the world. I wish he could have known how dearly loved and respected he was by millions of people he had never met. I wish he had known he would be interred in Westminster Abbey, between two of his scientific heroes, Isaac Newton and Charles Darwin, and that as he was laid to rest in the earth his voice would be beamed by a radio telescope towards a black hole.

But he would also have wondered what all the fuss was about. He was a surprisingly modest man who, while adoring the limelight, seemed baffled by his own fame. One phrase in this book jumped off the page at me as summing up his attitude to himself: 'if I have made a contribution.' He is the only person who would have added the 'if' to that sentence. I think everyone else felt pretty sure he had.

And what a contribution it is. Both in the overarching

grandeur of his work in cosmology, exploring the structure and origins of the universe itself and in his completely human bravery and humour in the face of his challenges. He found a way to reach beyond the limits of knowledge while surpassing the limits of endurance at the same time. I believe it was this combination which made him so iconic yet also so reachable, so accessible. He suffered but he persevered. It was effortful for him to communicate – but he made that effort, constantly adapting his equipment as he further lost mobility. He selected his words precisely so that they would have maximum impact when spoken in that flat electronic voice which became so oddly expressive when used by him. When he spoke, people listened, whether it was his views on the NHS or on the expansion of the universe, never losing an opportunity to include a joke, delivered in the most deadpan fashion but with a knowing twinkle in his eyes.

My father was also a family man, a fact lost on most people until the film *The Theory of Everything* came out in 2014. It certainly was not usual, in the 1970s, to find a disabled person who had a spouse and children of his own nor one with such a strong sense of autonomy and independence. As a small child, I intensely disliked the way strangers felt free to stare at us, sometimes with open mouths, as my father piloted his wheelchair at insane speeds through Cambridge, accompanied by two mop-haired blond children, often running alongside while trying to eat an ice cream. I thought it was incredibly rude. I used to try to stare back but I don't think my indignation ever hit the target, especially not from a childish face smeared with melted lolly.

It wasn't, by any stretch of the imagination, a normal childhood. I knew that – and yet at the same time I didn't. I thought it was perfectly normal to ask grown-ups lots of challenging questions because this is what we did at home. It was only when I allegedly reduced a vicar to tears with my close examination of his proof of the existence of God that it started to dawn on me that this was unexpected.

As a child, I didn't think of myself as the questioning type – I believed that was my elder brother, who in the manner of elder brothers outsmarted me at every turn (and indeed still does). I remember one family holiday – which, like so many family holidays, mysteriously coincided with an overseas physics conference. My brother and I attended some of the lectures – presumably to give my mother a break from her wraparound caring duties. In those days, physics lectures were not popular and definitely not for kids. I sat there, doodling on my notepad, but my brother put his skinny little-boy arm in the air and asked a question of the distinguished academic presenter while my father glowed with pride.

I am often asked, 'What is it like to be Stephen Hawking's daughter?' and inevitably, there is no brief answer that fits the bill. I can say that the highs were very high, the lows were profound and that in between existed a place which we used to call 'normal – for us', an acceptance as adults that what we found normal wouldn't count as such for anyone else. As time dulls the raw grief, I have reflected that it could take me for ever to process our experiences. In a way, I'm not even sure I want to. Sometimes, I just want to hold on to the last words my father said to me, that I had been a lovely daughter and that I should be unafraid. I will never be as brave as him – I'm not by nature a particularly courageous person – but he showed me that I could try. And that trying might turn out to be the most important part of courage.

My father never gave up, he never shied away from the fight. At the age of seventy-five, completely paralysed and able to move only a few facial muscles, he still got up every day, put on a suit and went to work. He had stuff to do and was not going to let a few trivialities get in his way. But I have to say, had he known about the police motorcycle outriders who were present at his funeral, he would have requested them each day to navigate him through the morning traffic from his home in Cambridge to his office.

Happily, he did know about this book. It was one

of the projects he worked on in what would turn out to be his last year on Earth. His idea was to bring his contemporary writings together into one volume. Like so many things that have happened since he died, I wish he could have seen the final version. I think he would have been very proud of this book and even he might have had to admit, in the end, that he had made a contribution after all.

Lucy Hawking

July 2018

Acknowledgements

The Stephen Hawking Estate would like to thank Kip Thorne, Eddie Redmayne, Paul Davies, Seth Shostak, Dame Stephanie Shirley, Tom Nabarro, Martin Rees, Malcolm Perry, Paul Shellard, Robert Kirby, Nick Davies, Kate Craigie, Chris Simms, Doug Abrams, Jennifer Hershey, Anne Speyer, Anthea Bain, Jonathan Wood, Elizabeth Forrester, Yuri Milner, Thomas Hertog, Ma Hauteng, Ben Bowie and Fay Dowker for their help in compiling this book.

Stephen Hawking was well known for his scientific and creative collaborations throughout his career, from working with colleagues on ground-breaking science papers to collaborating with script writers, such as the team from *The Simpsons*. In his later years, Stephen needed increasing levels of support from those around him both technically and in terms of communication assistance. The Estate would like to thank all those who helped Stephen to keep communicating with the world.

INDEX

Index

Very Brief Answers
to More Big Questions

Does it feel like a huge responsibility to have people expecting you to have all the answers to life's mysteries?

I certainly don't have the answers to all of life's questions. While physics and mathematics can tell us about how the universe began, they are not much use in predicting human behaviour because there are far too many equations to solve. I'm no better than anyone else at knowing what makes people tick, particularly women.

●

Among all scientists, who inspired you most, and why?

Galileo and Einstein. Galileo was the first modern scientist who realised the importance of observation, and Einstein was the greatest but, reassuringly, he had a number of blind spots, like quantum mechanics and gravitational collapse.

What do you think are the most important unasked or unexplored questions by humans today?

I can only echo Einstein's comment: the most incomprehensible thing about the world is that it is comprehensible. Why do we find ourselves in a vast and complicated universe that obeys laws that we can discover and understand?

•

How do you explain that, from time to time, the general public needs a kind of scientific icon (Einstein in his time, you now)?

Comparisons with Einstein are just media hype. Among physicists, I'm respected, I hope. But I'm just one of a number of people who have helped shape our modern view of the universe. We all need to know where we belong, and where we came from. But I'm not a genius, like Einstein was.

•

Has your disability influenced the way you do science?

My disease has hardly affected my work. That is because I was lucky enough to choose theoretical physics, which is one of the fields in which amyotrophic lateral sclerosis is not a serious handicap. I wouldn't have been able to continue had I been working in almost any other field.

What is your message to other people with disabilities?

My advice to other disabled people would be: concentrate on things your disability doesn't prevent you doing well, and don't regret the things it interferes with. Don't be disabled in spirit, as well as physically.

●

How did you get your electronic voice?

When I had my tracheostomy in 1985 and lost my voice, I thought at first that I might get it back eventually. But in the meantime, I desperately needed a way to communicate. I couldn't write or use a keyboard. All I could do was raise my eyebrows when someone pointed to letters on an alphabet card. But then a software engineer called Walt Woltosz sent me an early Apple computer with the first version of the program I am using now. At first, people had to read the screen to see what I was saying. But about a year later, I got this voice. I have stuck with it, because it has become my trademark. People sometimes poke fun by imitating it, but I regard that as a sign of success. It is like a politician have a *Spitting Image* puppet.

●

How do you prepare your lectures?

I use the blink switch and write at around five words a minute.

Why is Galileo said to be the father of modern science?

Galileo was the first person to challenge the received wisdom that the ancient Greeks, and Aristotle in particular, were the ultimate authority in science. Galileo pointed out that simple observations, like dropping weights from a height, show things do not work the way Aristotle said. This must have been seen by many people but they had put it down to imperfect observations, or other reasons. But Galileo said the ancients were actually wrong, and started to work out the correct laws from the observations. That makes him the father of modern science.

•

Isaac Newton was born on Christmas Day 300 years before you were born, and you now hold the same position he did at Cambridge. What affinity, if any, for Newton do you feel given these similarities?

Perhaps Newton's greatest achievement was providing a single theory that presented a unified explanation of a wide range of what had previously been thought to be disparate phenomena. Newton's Laws of Motion and Gravitation not only explained the movement of bodies in the heavens and billiard balls on Earth, they also explained the rise and fall of the tides.

Like Newton, I believe that scientists need to find a single unifying theory to explain the large and the small, the large-scale structure of the cosmos as well as the interior of the atomic nucleus. After all, the Big Bang theory supposes that when the universe began it was infinitesimally small.

•

If we could get Newton up to speed on everything we know today, and if you could speak to him through time, what problem of today would you task him with solving?

Is the solar system stable? And what happens to a star that cannot support itself against its own gravity?

•

What was your craziest revelation ever?

My idea that seemed craziest at first sight is that black holes are not completely black, but glow like hot bodies, with a temperature that is higher the smaller the black hole.

•

Black holes have proved so enduringly interesting – to you, to other scientists and also to the general public. Can you describe, briefly, what they are?

Black holes are quite literally holes in space that stuff can fall into, but not get out of. They are places where the gravitational field is so strong that nothing, not even light, can get away. They are formed by the collapse of massive stars when they have exhausted their nuclear fuel, and can no longer support themselves against their own gravity. The boundary of a black hole is called the horizon.

●

Is there a possibility that a black hole might consume Earth or its neighbouring planets in the near future?

No, it's not something you have to worry about any time soon. Even our nearest black hole is 2,800 light years away.

●

You suggest humanity is at risk because of several factors such as global warming, genetically engineered viruses and war. Aside from establishing self-sustaining colonies in space, what major piece of advice would you give us and generations to come to make the world safer and better prepared?

It's a bit like trying to put the genie back in the bottle after it has been released. The knowledge of how to make nuclear weapons is available, and the material to make them can be obtained by a determined organisation or country. The only suggestion I can make on the danger that nuclear weapons present is that we should put a lot of effort into developing nuclear fusion rather than using the nuclear fission currently relied on for power and bombs, and lock all fissile material in deep bore holes. But it won't save us from other technologies and their dangers. The only way to avoid these dangers would be some form of world government. But that might become a tyranny and is not the way that the world is currently moving with Brexit and Trump.

●

We live in times of extraordinary technological advances and scientific progress. How do you explain the current anti-science sentiment, from climate change denial to the anti-vaccination movement?

I agree. People distrust science because they don't understand how it works. It seems as if we are now living in a time in which science and scientists are in danger of being held in low, and decreasing, esteem. This could have serious consequences. I am not sure why this should be, because our society is increasingly governed by science and technology, yet fewer young people seem to want to take up science as a career. Perhaps new and ambitious space programmes will excite them and stimulate interest in other areas such as astrophysics and cosmology.

In 2016, you announced a project named Breakthrough Starshot with the goal of sending a fleet of interstellar nano-spacecraft with sails to Alpha Centauri, our nearest neighbouring star system, in search of life. Can you outline the technology being built for these probes and the odds of the probes reaching their destination? And what do you hope to find?

The nanocraft are pushed to 20 per cent of the speed of light by giant lasers on Earth. Nanotechnology has advanced in recent years and will allow us to send back pictures and do other tests for life. The chances of success are slim but are weighted against the importance of the discovery of alien life on our nearest neighbour, and make the project worthwhile.

•

Was your experience of a zero-gravity flight all that you thought it would be? What do you think is the future of human space flight?

The flight was amazing. The zero-G part was wonderful, and the high-G part was no problem. I could have gone on and on. After decades of disinterest, space travel is finally coming back into the news.

I think the survival of the human race will depend on its ability to find new homes elsewhere in the universe, because there's an increasing risk that a disaster will destroy Earth. I therefore want to raise public awareness for space flight. A zero-gravity flight was the first step towards this.

Photograph © Steve Boxall/steveboxall.com

Can you describe why Earth works for human life now? What are the conditions that allow human life to live here on Earth?

Life on Earth is possible only because a number of parameters lie in certain ranges. Some of these are clearly environmental, like the Earth has the right temperature and pressure to have liquid water. In a galaxy of a hundred billion stars, many of which have several planets, these conditions will be met somewhere. But the facts that galaxies and stars exist at all, that chemical elements are created by nuclear fusion reactions inside stars, and that chemistry allows complicated molecules, depend on the parameters of the theory lying in a small region. Is this apparent fine-tuning of the constants of nature evidence of design, or is it another environmental effect? According to M-theory, our best candidate for a theory of everything, the same fundamental theory can lead to a large number of effective theories, with different values of the parameters, the so-called constants of nature. Somewhere in this vast landscape of effective theories, there should be some that allow life on planets like Earth. Let's hope there are such theories, or we don't exist.

If you were to tell aliens about the greatest achievements of our human civilisation on the back of one envelope, what would you write?

Gödel's incompleteness theorems and Fermat's Last Theorem are things aliens would understand. But maybe this is too narrow. However, it is no good writing about beauty, music or poetry because they are human-specific. Aliens would not understand these concepts.

●

We hear you are a big *Star Trek* fan, even appearing in an episode as a hologram. Over the span of two *Star Trek: The Next Generation* episodes – 'Elementary, Dear Data' and 'Ship in a Bottle' – they instruct the computer to create a unique Sherlock Holmes mystery with an adversary capable of defeating Commander Data. The adversary, Professor Moriarty, is made aware that the holodeck programme and he are simulations. The crew eventually trick Moriarty and 'beam' him off the holodeck, into another simulated reality – and they set him off on a lifetime of exploration and adventure. Reality may actually be a fabrication generated by 'a little device sitting on someone's table'. What is the evidence we are not in a simulation? How could we even prove it?

Philosophers from Plato onwards have argued about the nature of reality. According to classical science, there exists a real external world whose properties are definite and independent of us, the observer who perceives it. Both observer and observed are parts of a world that has an objective existence. In science fiction, however, a different kind of reality is exposed. In the film *The Matrix*, people live unknowingly in a simulated virtual reality created by intelligent computers. And maybe this isn't so far-fetched. Many people prefer to spend their time in the simulated reality of online virtual worlds such as 'Second Life'. How do we know that we are not just characters in a computer-generated soap opera? It would be lawless, and we just playthings. As long as the simulation obeys a set of consistent laws, we cannot distinguish it from reality. And there is no way we could prove there was another reality behind the simulated one. It then becomes our reality.

●

If you had an opportunity to travel into the past or the future, which era would you choose, and why?

I would travel to the future. We already know the past from history.

●

Which living person do you most admire, and why?

A few years ago, I would have said Nelson Mandela. He brought a peaceful solution to a seemingly impossible situation. Now that there is no one on the political scene of his stature, I would say the physicist Edward Witten. He is not known widely outside academic circles, but his work has been an inspiration to me and to other physicists. He is currently working in Princeton and is researching the somewhat mystifying topics of string theory and quantum gravity. His importance is probably demonstrated by his being the only physicist to be awarded the Fields Medal for mathematics. This is a hugely significant acknowledgement of Witten's ability. As was said when he received the medal, 'Although [Witten] is definitely a physicist . . . his command of mathematics is rivalled by few mathematicians. Time and again he has surprised the mathematical community by a brilliant application of physical insight leading to new and deep mathematical theorems . . . he has made a profound impact on contemporary mathematics.'

•

How would you like the world to remember you?

I would hope I would be remembered for my work on black holes, and the origin of the universe. But above all, I would like to be remembered by my children and grandchildren, as a great dad and grandad.